Praise for *The Invisible Machine*

"The nervous system can be reset. The significance of that, recognized by the legal system and by the medical profession, would be revolutionary."

—Gabor Maté, MD, *New York Times* bestselling author of *In the Realm of Hungry Ghosts* and *The Myth of Normal*

"In *The Invisible Machine*, Lipov and Mustard rewrite the recovery narrative for many who have experienced personal disruptions and life compromises. Through personal stories we learn that traumatic experiences may lock the nervous system into a chronic state of defense and fight-flight behaviors, while compromising feelings of safety and sociality. The Lipov treatment gives permission to retune, enabling one to return to sociality, playfulness, and social connection."

—Stephen W. Porges, PhD, author of *The Polyvagal Theory*, founding director of the Traumatic Stress Research Consortium at Kinsey Institute Indiana University Bloomington, and professor at University of North Carolina at Chapel Hill

"*The Invisible Machine* is a powerful book, filled with hope on recovery from the trauma that haunts millions of people. I highly recommend it."

—Daniel G. Amen, MD, founder of Amen Clinics and *New York Times* bestselling author of *The End of Mental Illness* and *Change Your Brain Every Day*

"*The Invisible Machine* combines Lipov's astonishing innovation with Mustard's keen analysis and masterful integration. This book is a compelling game-changer."

—John "Jay" Faber, MD, double board-certified forensic trauma specialist and former national medical director for Humana's behavioral health division

"Dr. Lipov discovered and perfected a modern treatment for post-traumatic stress. He understands how my patients hurt and has helped them heal. His work with Jamie Mustard is an amplifier of change. This book rewrites the narrative of trauma as a physical *injury* and is a must read."

—Frank Ochberg, MD, founding father of trauma science and doctor who coined the terms "Stockholm syndrome" and "post-traumatic stress injury"

THE INVISIBLE MACHINE

Also by Eugene Lipov

Captain Heart

Also by Jamie Mustard

The Iconist

Also by Holly Lorincz

Crown Heights
Arsenal of Hope

THE INVISIBLE MACHINE

The Startling Truth About Trauma and the Scientific Breakthrough That Can Transform Your Life

EUGENE LIPOV, MD

JAMIE MUSTARD

HOLLY LORINCZ

BenBella Books, Inc.

Dallas, TX

This book is for informational purposes only. It is not intended to serve as a substitute for professional medical advice. The author and publisher specifically disclaim any and all liability arising directly or indirectly from the use of any information contained in this book. A healthcare professional should be consulted regarding your specific medical situation. Any product mentioned in this book does not imply endorsement of that product by the author or publisher.

Proceeds from the sale of each book will be donated to Erase PTSD Now.

BenBella Books, Inc.
10440 N. Central Expressway
Suite 800
Dallas, TX 75231
benbellabooks.com
Send feedback to feedback@benbellabooks.com

BenBella is a federally registered trademark.

Printed in the United States of America
10 9 8 7 6 5 4 3 2 1

Library of Congress Control Number: 2022041575
ISBN 9781637741603 (hardcover)
ISBN 9781637741610 (electronic)

Editing by Gregory Newton Brown
Copyediting by Scott Calamar
Proofreading by Marissa Wold Uhrina and Ashley Casteel
Indexing by Amy Murphy
Text design and composition by PerfecType, Nashville, TN
Cover design by Pete Garceau
Cover image © iStock / duncan1890 (human brain) and / dem10 (MRI scan)
Printed by Lake Book Manufacturing

For my grandmother Dorothy Gilmer Ross, who gave unconditional love.
For Corey Drayton and his genius, stewardship, and artistry.
And Trevor Beaman, for his courage, friendship, and brotherhood.

—Jamie Mustard

To my wife, Robbin, for her strength and support; my son, Sam, for his humor; my mother for her warmth and empathy; my father for his strength and determination; and my brother, Sergei, for his intelligence and vision.

—Dr. Eugene Lipov

CONTENTS

FOREWORD

Post-traumatic stress (PTS) is a natural, normal, temporary response to a sudden, dehumanizing event. We humans are humane because we suffer and empathize with the suffering of others. Post-traumatic stress helps our species survive.

My generation recognized post-traumatic stress as morally and medically normal, yet defined post-traumatic stress *disorder* (PTSD) as a mental illness. We modified that definition over the decades, as our science and our understanding evolved. In 1980, we failed to realize that the term *disorder* was both stigmatizing and inaccurate. When a survivor of severe trauma and traumatic stress cannot recover within a month, that person deserves the accuracy and the dignity of the term post-traumatic stress *injury.*

This injury is physical. The results are medical as well as emotional. Like how scalding heat harms our skin, a searing event changes our neuroanatomy and our neurophysiology. Fifty years ago, we could not detect these changes through routine neuroimaging. Now we can. These visible changes affect our thoughts, actions, and reactions. Post-traumatic stress symptoms are a natural, biological response, built into our DNA as a survival mechanism—one that can be turned on but not always turned off. This is one of the major reasons I want to see the term *PTSD* changed to *PTSI.* It's a physical injury, not a disorder. That means post-traumatic stress injuries can be healed, the same as burns and broken bones. The wound is just hidden within our nervous system and brain.

PTSI is a biological imperative that must be critically examined by every sector of our society. It is like a weed with a deep root system we can't seem to eliminate. Could it be possible that preventable forms of crime, mental illness, discrimination, child abuse, mass shootings, homelessness, and disease have proliferated because of post-traumatic stress *injury*?

All these years, I have asked, "How can we preserve human nature *and* reduce human cruelty?" One compelling answer to that question is in this book, *The Invisible Machine*, with the latest research and groundbreaking science from colleagues who have advanced the treatment of post-traumatic stress injury. There is hope.

—Frank Ochberg

World-renowned trauma specialist who defined the term *Stockholm syndrome* and who, along with Dr. Jonathan Shay, coined the term *post-traumatic stress injury*

INTRODUCTION

Joe stepped through the door onto his front stoop, took a sip of coffee from his favorite travel mug, and tilted his face toward the morning sun.

A loud blast from the neighbor's car horn disrupted the peace. Joe's body twitched, hard, and his heart leaped into a fast beat. His cheeks flushed. Swiping at coffee on his jacket, he loudly cursed the neighbor.

After his cathartic outburst, Joe should have settled down within seconds, been able to shake it off. Instead, the curse words continued to flow as he punted a small toy off the sidewalk and stalked to his car, wrenching the door open.

Have you ever overreacted this way? Probably yes. We all have our moments when we let little things get to us. However, some of us respond this way multiple times a day, every day, with over-the-top reactions at inappropriate times.

Do you know why?

It's because you're human, and you've experienced trauma.

Despite what you have heard in the movies, post-traumatic stress injury (a biological phenomenon commonly and wrongfully referred to as *post-traumatic stress disorder*) isn't just for those who have survived a war or brutal abuse. Every level of trauma causes stress to the body to some degree, no matter whether you are a grocery store clerk or a combat medic.

Discussed in detail in these pages, you'll hear about how the sympathetic nervous system (dubbed "the invisible machine" by the authors) can be triggered into a consistent state of overactivation by big trauma, but also by lesser amounts of stress occurring over a long period. And this overactivation, though invisible to the naked eye, can now be captured by scans.

The scans reveal what an overactive sympathetic nervous system does to the part of the brain called the *amygdala*, creating a feedback loop that keeps the "fight-or-flight" mode activated even when danger is long gone. This permanent state of fight or flight leads to post-traumatic stress and symptoms revolving around anxiety, insomnia, nightmares, sense of doom, hair-trigger responses, paranoia, anger, and hypervigilance.

Anybody can acquire trauma that leads to PTSI. And now a reset exists that can take away the misery.

The goal of this book is to provide authentic, observable change in how we take on post-traumatic stress as a society. The authors offer a contemporary understanding of post-traumatic stress injury and how many, many more people suffer from post-traumatic stress than is generally acknowledged; how a quick, accessible biological intervention exists and has been proven safe and far more effective than other healing modalities; and how the proliferation of post-traumatic stress unnecessarily impacts society as a whole.

You may be asking yourself: Who are we to tell you what to think? Well, first of all, I am simply a recorder, collaborator, and word honer. I was asked to join the team because of my experience writing about PTSI and criminal justice. I worked with *The Invisible Machine* team to help capture and refine the research, the interactions, and the genius intuitive leaps that the authors have shared with me.

On the other hand, author Eugene Lipov, MD, a board-certified physician in anesthesiology and pain, is one of the world's leading experts on the physical consequences of post-traumatic stress. Based out of Chicago, he is the chief medical officer of an ever-growing number of Stella Centers, clinics that offer his innovative treatment for PTSI. Eugene pioneered a simple, safe procedure (based on a century-old pain anesthetic procedure) that successfully treats trauma. Years of clinical documentation on thousands of patients demonstrate an efficacy rate upward of 85 percent, or more, with minimal side effects. His treatment is touted by his most prestigious peers, including Dr. Frank Ochberg, Dr. Daniel Amen, Dr. Jay Faber, and Dr. Stephen Porges. Multiple peer-reviewed studies have verified the safety and efficacy of the treatment. NYU is currently conducting a $3.8 million research study based on Eugene's procedure. This procedure was endorsed by Barack Obama, and it has been featured in many major media outlets, including *60 Minutes*, *CBS This Morning*, and the *Joe Rogan Experience*.

Author Jamie Mustard is an artist, a strategic multimedia consultant, product futurist, and thought leader. A graduate of the London School of Economics, he is best known for his work with high-profile companies to make their ideas or products stand out, and for his book *The Iconist* (2019 Outstanding Works of Literature award winner) based on his "economics of attention" principle. As a child, he suffered years of sustained high- and low-level trauma, and it was only after stumbling across Dr. Eugene Lipov and his procedure that he found relief from PTSI.

Since that day, Jamie has brought together a wide array of people to help promote Eugene and his life-changing innovation, from Special Forces operators and earthquake survivors, to the incarcerated, to famous doctors, to everyday people suffering from the effects of trauma. How the universe placed each influential individual in Jamie's path is fascinating and seems a minor miracle.

With so many PTSD/PTSI books on the market, you may be asking: What makes this one noteworthy? It's because we talk about the largely undiscussed underlying issues around trauma *and* offer an actionable solution without dependency. *The Invisible Machine* also addresses the medical industry that has been slow to take notice and provides an in-depth analysis, which includes confronting the critics' arguments, of Eugene's procedure.

When the threat of post-traumatic stress to individuals goes unacknowledged, the impact of the symptoms is harmful and far-reaching. The many studies and decades of clinical data conclusively reveal long-term mental and physical issues caused by post-traumatic stress. Years of doctor visits and prescriptions tie up the nation's healthcare system and have done little to stem the tide of harmful symptoms for most people. Then there are the statistics on PTSI-based suicide, which are devastating.

These individual struggles have broader implications for society as a whole—the society *you* live in. While the authors assess the similarities between all cases of PTSI, no matter how or where it is acquired, there is also a focus on the parallel nature of post-traumatic stress in combat soldiers and in criminals, and how both respond when their nervous systems are "reset" and the symptoms are reversed. Reintegration into family and community is an issue for returning combat soldiers and those just released from prison; the culture shock is a lot easier to maneuver if the person isn't dealing with PTSI symptoms and can remain grounded and able to fully function.

Imagine a world where recidivism rates drop. Or even being able to stop criminal behavior before it advances. A world where our soldiers can return home and feel at peace and restore healthy relationships. A world where people who've experienced years of abuse or those who have gone through a long, slow drip of micro-traumatic experiences can shut down all the symptoms of PTSI and move on with healing emotionally.

What if that world isn't as far off as it seems? What if Eugene's procedure can get us there, as it has for the many voices in the coming pages?

With less pain in the world, we could see less stress in everyday households, leading to healthier (physically and emotionally) children and adults, to happier and healthier work environments, less crime . . . If that sounds too good to be true, have no fear; this book is realistic and data driven. We are not dreamy idealists selling a miracle cure. We don't think the procedure you will learn about will cure humans of being human. Frankly, we don't use the term *cure* at all. It is a reset.

TAKING ON THE TRUTH

Real, tangible relief now exists with a diagnosis of post-traumatic stress. Eugene's procedure is biological, it is safe, it is accessible, and it has a clinical success rate and longevity that far outstrips other treatments. This is because many approaches take on the external symptoms of PTSI while Eugene's biological innovation treats the root of the physical problem—the overactivation of the sympathetic nervous system, or what he calls the overactivation of the invisible machine.

The procedure, now called the *dual sympathetic reset* (DSR), is a highly innovative version of a century-old nerve block called the *stellate ganglion block* (SGB). DSR is simple: a double injection of an ordinary, local anesthetic on the right side of the neck at the C3/C6 vertebrae, which is then repeated on the left side, if necessary.

"I believe this might just be the most impactful medical innovation since the advent of penicillin in 1928. And I intend to prove it," says Jamie. "This will change the way humans move through the world. As a human step, I compare it to the moon landing in 1969."

A random conversation set this medical innovator and artist on a scientific, humanity-oriented journey to bring the procedure to the people. Along the way, the universe pulled a handful of exceptional strangers into the quest, forging a small band of strong-willed characters who have joined the good fight, making sure change happens for individuals and around the globe. This group of intense strangers (Jamie's creative partner likes to call them "The Avengers") has come together to support Eugene's work becoming a new normal for PTSI treatment. Some influencers you will meet in these pages: a war correspondent, a female sheriff presiding over one of the largest jails in the United States, and several Special Forces officers. They all want to stop unnecessary suffering in others.

The interviews used in *The Invisible Machine* were conducted by Jamie and his team, while the medical knowledge and data collection comes from Eugene. Jamie and Eugene shaped this content, provided and explained research findings, and offered hundreds of patient anecdotal evidence, stories, interviews, and articles.

This is wholly their book. The honor has been all mine.

—Holly Lorincz, collaborative writer

Note: While referencing post-traumatic stress symptoms, the more accurate term *PTSI* is used most of the time. However, the more common term *PTSD* is used when it is part of a quote or summary from source material or if it fits that particular content's timeline.

CHAPTER ONE

The Tiger and You

In an upscale Mexican restaurant in downtown Chicago, the windows were thrown open to the damp night due to COVID-19. Artist Jamie Mustard listened to Dr. Eugene Lipov explore the topic of trauma over platters of stuffed chiles and beef birria tacos.

In his light Ukrainian accent, Eugene asked Jamie, "If you were a caveman forced to run from a tiger all the time, what would that do to you?"

"I'm guessing nothing good?" Jamie answered. If anyone was going to know about the consequences of large or long-term doses of stress, it was the premier pain physician and board-certified anesthesiologist across from him.

"That is true," Eugene said. "Thirty thousand years ago, we were surrounded by constant threats and had to stay alert or die. Which, obviously, is traumatic. That hyperalert response to stress or danger is *biological.* It is a physical reaction built into the DNA that can make us miserable in the long run but keeps us alive."

The doctor talked about how the reality of a large, aggressive animal bursting into a caveman's shelter was a genuine, imminent, and persistent threat. That caveman had little to rely on to keep himself alive except his ability to respond quickly, to remain in what is popularly referred to as fight-or-flight mode.

Pausing briefly to order a bottle of wine, the doctor continued. "If you didn't want to die, you'd need to have a hair-trigger response, be ready to fight at any moment. You wouldn't sleep because, you know, there's a tiger hanging around. Have to stay alert, be at the ready. This is hypervigilance. You're going to have a hair-trigger response to any stimuli. You're going to have extreme paranoia and anxiety, knowing that a big cat is likely around the corner, licking his chops.

"And you'd be contemplating your death all the time. Homicidal or suicidal ideation, thanks to fearing that tiger's attack day in and day out—you'll eventually just want it to be over with." Eugene took a sip of his wine, then said, "And you're going to have deep erectile dysfunction. Because if you are running from a tiger, you're not able to have sex."

Jamie grimaced, then chuckled. A thought leader according to *Forbes*,[1] he was not easy to impress and, yet, he found himself sucked into Eugene's storytelling. The doctor had a way of talking that was educational while entertaining, reminding Jamie of his favorite professors at London School of Economics. A leader in the emerging field of psycho-anesthesia, Eugene was the chief medical officer at the Stella Center,[2] known for his use of complex anesthetics during heart transplants or other complex trauma surgeries and for successfully treating nerve pain; innovations in the treatment of hot flashes; and, now, pioneering a physical relief to post-traumatic stress symptoms.

The symptoms that had haunted the overly vigilant, stressed-out caveman were the same kind of issues that had brought Jamie to Dr. Lipov's clinic earlier that day.

"Millions and millions of people around the world may be dealing with long-term trauma fallout," Eugene told Jamie, "whether they recognize it in themselves or not. But, as you've discovered, I've found a way to help them."

Leaving dinner that night, Jamie found that Eugene's statement impacted him deeply. It would drive his thoughts for months. It became a call to action.

WHAT HAS TRAUMA DONE TO YOU?

The two people described below seem to have nothing in common. But they do. They both have endured trauma. They both suffer from post-traumatic stress.

A business owner, young and physically fit, loses consciousness. From afar, he hears the ER doctor say, "He's crashing. Code Blue!" When he wakes, he is told his heart stopped briefly. Why? He's shocked to find he's suffered a severe panic attack.

An Army Reservist wakes from a heavy sleep to hear his wife grunting. She is beating at his chest—he has his hands wrapped around her neck, choking her so hard that she is being lifted from the bed. This is the third time he has awakened to find he is trying to kill the woman he loves.

Frontline workers, abuse survivors, and disaster victims dealing with post-traumatic stress are discussed widely in today's medical journals, as well as in mainstream movies and television shows. No one finds it surprising any longer that extreme trauma can lead to post-traumatic stress symptoms. However, what is not discussed is how many *everyday people* live with those same symptoms—and how often those everyday people do not recognize post-traumatic stress symptoms in themselves.

According to the National Council of Mental Wellbeing, 70 percent of adults have experienced at least one impactful traumatic event in their lifetime in United States. That's 223.4 million people.[3] That's twenty-eight out of the forty people living on your street.

The CDC states that 20 percent of people who experience a traumatic event will develop what has commonly been labeled PTSD, which means some of your neighbors are currently dealing with symptoms like hypervigilance, erratic anger, insomnia, as well as various health conditions like heart disease.[4] Or maybe it's you.

We propose that the number of those suffering from post-traumatic stress is actually closer to 30 percent—possibly up to 60 percent. At this point in history, these numbers are swiftly trending upward, thanks to COVID, social upheaval, and now a large-scale war in Europe. As it is, trauma experiences and their consequential health symptoms are vastly underreported, if acknowledged at all.

And why is that?

When you hear the term *PTSD*, your brain likely conjures up a scene from a popular movie with a man in fatigues suffering a breakdown. Post-traumatic stress disorder is a widely recognized diagnosis and accepted health issue for combat soldiers, as the military has a population that experiences obvious trauma. Furthermore, soldiers with trauma symptoms are easier to track and treat; they are part of a highly regimented organization with vast records. The military also has the means to research and fight post-traumatic stress symptoms and has found some success—including using Eugene's innovation on a large scale.

The problem is that the focus has been solely on the military community for so long. A PTSD diagnosis has become linked with extreme trauma, even heroism, in the minds of the populace. And that is *far, far from the truth*. Everyday individuals are suffering with no clue as to why, and it is causing a public health crisis within the healthcare system and the criminal justice system.

It *is* public knowledge that being involved in big, traumatic events like 9/11 can lead to post-traumatic stress, but in order to understand how so many of us can acquire post-traumatic stress symptoms without realizing it, the full definition of trauma must be better broadcast.

We will be delving into trauma more thoroughly throughout the book, but know for now that the average person with unrecognized trauma often has dealt with microdoses of stress over a long period (for example, growing up with food insecurity or with neglectful parents). That stress can build up and reach a point at which the person develops the same post-traumatic stress symptoms as a Navy SEAL, an assault victim, or a 9/11 survivor.

The truth is, trauma—no matter how it is inflicted and no matter how "big" or "small" we perceive the trauma to be—can create a cascade of biological changes that produce an overactive sympathetic nervous system. This forces the brain's amygdala (the part of the brain that in large part controls the stress response) to switch on the fight-or-flight response and keep it on, creating a feedback loop even when the danger or stress is long gone.

The Sympathetic Effect of Threat

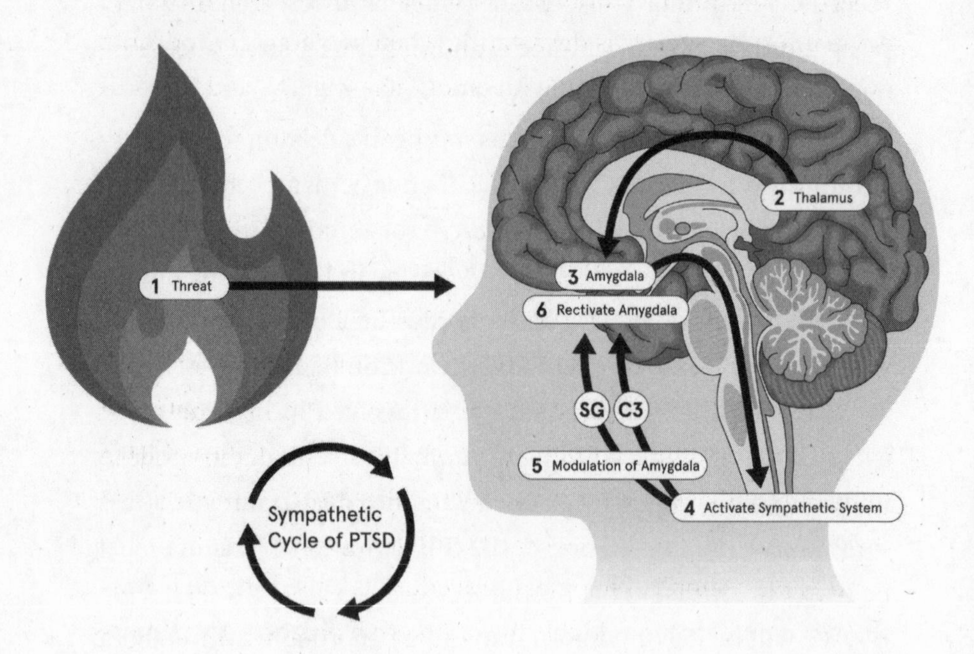

The threat is perceived through the senses; the thalamus confirms the threat; the amygdala reacts, activating fight or flight; the sympathetic nerves carry the message throughout the body and back into the amygdala.

This overactivation of the sympathetic nervous system is invisible to the naked eye but *visible on an advanced neuro scan*. Overactivation is a physical, measurable change and is the cause of post-traumatic stress symptoms.

Hence, the more accurate diagnosis is post-traumatic stress injury. This is a biological injury, not a disorder.

PTSI is *hugely* underdiagnosed and can affect anyone from a yoga instructor, to an inmate, to a CEO. The broader, trickle-down effect on society is devastating when you consider the large population of us struggling with emotional, mental, and physical issues thanks to PTSI.

RELIEF

Both the soldier and the business owner described earlier suffered from debilitating symptoms of post-traumatic stress—that is, until Eugene was able to reset their sympathetic nervous system in a ten-minute outpatient procedure at his clinic. That's right: no matter the level of trauma, *an overactive fight-or-flight response can be reversed back to a pre-trauma state.*

Eugene's simple, biological procedure is called the *dual sympathetic reset.* DSR is a double injection of an ordinary local anesthetic on the right side of the neck at the C3/C6 vertebrae (and on the left, if necessary)—a miraculous innovation based on a safe, century-old nerve block called the stellate ganglion block (SGB). After Eugene's adaptions, "comparing SGB to DSR is like comparing a sleek, new motorcycle to a basic child's bicycle," according to Jamie.

The DSR procedure is now available to civilians and soldiers alike; a large investment firm has opened Stella Centers around the United States. Eugene, based at the Chicago Stella Center, serves as the chief medical officer, personally training each physician.[5] The cost is roughly $3,000. There are a few grant programs to help those who can't afford the treatment (see the final chapter).

To date, somewhere between five and ten thousand people with post-traumatic stress have been successfully treated with SGB, or the now updated DSR version (which is currently part of a massive clinical study at NYU). Numbers are hard to pin down, since many of the procedures are done at military institutions and are underreported. Clinical data reports few side effects,[6] an efficacy rate of 80 to 85 percent, and a number of patients now reaching ten to fifteen years of relief. For those who are retriggered, getting the outpatient procedure done a second or third time usually results in relief from the symptoms.

The complication rate is at an extremely low 1.7 per thousand stellate ganglion blocks,[7] according to the National Library of Medicine. These numbers are based on a large study by Dr. H. Wulf in Germany, with forty-five thousand SGB procedures (pre-DSR) performed before computer imaging was incorporated for further accuracy. Now that doctors use ultrasounds for needle placement, the complication rate is even lower.

The results are real. The relief is available now.

THE MANY FACES OF TRAUMA

The business owner who thought he was fine but ended up in the hospital because of a severe panic attack is a real person, a man who had lived with a steady stream of micro-traumas the bulk of his life. He underwent DSR, and within twenty-four hours he was able to experience peace and deep joy for the first time.

Walking away from the clinic, he told his wife, "I feel like I have a second chance at life. The pressure is gone. I feel like I can *feel* again."

Keep in mind, as we've said, these effects of post-traumatic stress can haunt anyone, not just those with big, crazy trauma. And DSR can help anyone on the spectrum of trauma sufferers.

The Army Reservist who tried to strangle his wife in his sleep is another real person. Jason Brown heard about DSR (though it was still called stellate ganglion block at the time) and decided to take a chance. Jason was terrified he was going to kill his wife, or, at the very least, she was going to leave him. He had returned to the United States from his last tour of combat duty in Iraq with severe PTSI symptoms, including paranoia, constant unease, and night terrors, and they were only getting worse.

Overseas, he'd been the daily target for snipers. The rock-bottom moment had come when he was forced to kill a ten-year-old child—the boy had run up to his unit on a city street, strapped into a bomb vest with his hand on the trigger. The soldier had to shoot the boy in order to protect his troops.

Back home, his mind was unable to make the switch from battle mode to civilian. He was suicidal, desperate for help before someone got hurt. He came to the Stella Clinic, where Dr. Lipov performed SGB.

Pre- and post-procedure brain scans revealed a dramatic change in his brain, the deactivation of the amygdala. Jason finally achieved relief. No more night terrors or sleep attacks.

Jamie has considered the term *trauma* many times since meeting Eugene and interviewing people like the business owner and the soldier.

Base State / No Trauma State

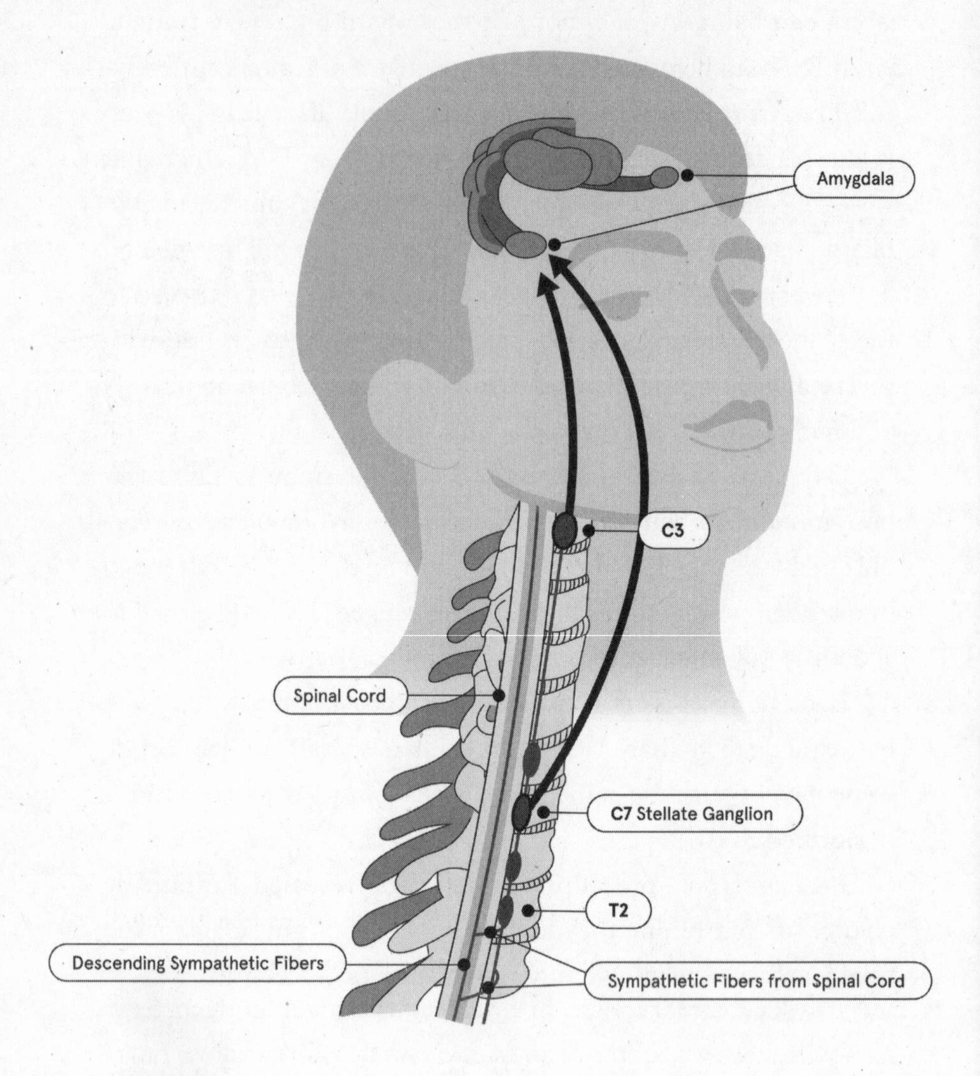

Trauma

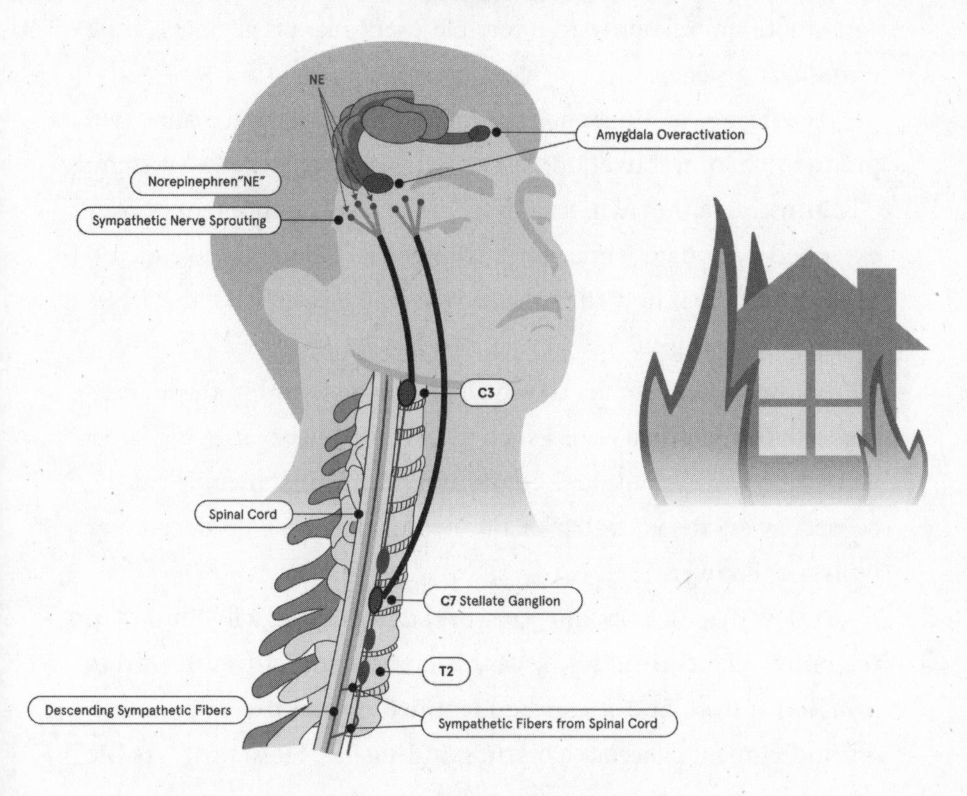

The artist likes to say that trauma, like love, is a primary driver of the human experience. It can drive humans to become successful artists, or CEOs, or a prime minister, but too much for too long can drive people to addiction and despair. Trauma hurts individuals, which in turn hurts society.

The *Oxford Dictionary* defines trauma as "a deeply distressing or disturbing experience," and an "emotional shock following a stressful event or a physical injury, which may be associated with physical shock . . ."[8]

The American Psychological Association describes trauma as "an emotional response to a terrible event like an accident, rape or natural disaster . . ."[9]

According to Eugene, trauma is any occurrence that will produce an overactivation to the sympathetic nervous system: "Trauma is trauma, whether it occurs in a single moment or over extended periods of time, even whether it is huge or considered minor. The key is how the trauma is perceived and what the DNA of the person is like."

As mentioned above, trauma can also be found in those dealing with less-defined issues such as chronic illness or living in an emotionless/overemotional environment. For instance, it can be caused by an absent father or mother, or parents who devalue or bully a child over time.

A slow drip of constant stressors such as these will "build" in the body. The clinical term for this cumulation of trauma-stress is *allostatic load*, and it can be identified by the use of biomarkers and clinical criteria, according to Eugene. He states, "As life happens and people endure stressful events, micro or macro, this stress load accumulates. An *allostatic overload* happens when an individual hits the tipping point and can no longer cope with the ensuing stress."

Trauma is anything that happens too soon, for too long, or too forcefully so that the mind, soul, and body cannot process the event and return to homeostasis (maintaining equilibrium)—whether the person recognizes the trauma or not.

Sadly, many people who've suffered trauma at a low level over a long time are unaware of their high allostatic load, often assuming the misery they deal with, emotionally or physically, is status quo.

THE INVISIBLE MACHINE

As Eugene and Jamie talked and ate together that first night, the pain physician sipped his wine and often broke the tension with a joke, an easy laugh, or a well-timed curse word while sharing his vast experience, education, and training around how trauma impacts the body.

"Reacting to trauma isn't necessarily bad. In everyday life, humans can endure emotional or physical trauma, big or small, and the sympathetic nervous system will be activated, giving the human the boost they need in the time of crisis. Then, optimally, they will return to baseline when the stress passes. The fight-or-flight response will be deactivated.

"However, if that danger (real or perceived) doesn't go away," said Eugene, "I like to say our invisible machine gets stuck in the 'on' position."

"What are you calling an invisible machine? Our sympathetic nervous system?" Jamie raised an eyebrow.

"Yes. Much of the sympathetic nervous system is exactly that, a machine made up of filaments invisible to the naked eye," said Eugene. "However, the stellate ganglion is the node where sympathetic nerves connect in the neck and it *is* visible to the eye. The nerve fibers leaving the stellate ganglia connect to the amygdala, and that is the bridge that allows them to talk . . . basically, a feedback loop which controls our fight-or-flight response. Like I've said, if the trauma is big enough or occurs over a long time, the sympathetic nervous system will overactivate, making the machine think the fight-or-flight mechanism needs to stay on. All the time. For some, this lasts for years. Possibly decades."

When there were tigers around every cave corner, the human species survived (even if the act of procreation may have not always been easy) by operating in a kind of permanent state of high alert, a biological imperative.

However, that same biological imperative no longer exists, not at a species level.

Jamie considered this. "The human race is not in jeopardy of dying out anymore, at least not from a rampant tiger threat," he said. "The world will go on if I let my guard down while cuddling on the couch with the family dog, eating Cheetos, and binge-watching *House Hunters*. The threat of a tiger, or any predator, jumping in through a window at any second is very low for most of us. Unless you live in a slum like the one I grew up in."

"Okay, yeah, but there are still dangers," said Eugene. "The fight-or-flight response is still necessary. Obviously, we need to be able to react quickly when danger is present or a traumatic event is in play. As you say, very few of us live in a situation where we need our invisible machine to stay on high alert at all times—and yet it does."

THE EVOLUTION OF A LABEL: DISORDER VERSUS INJURY

The human race has evolved, and unless a person is dealing with exceptional, horrific circumstances (war, long-term abuse), it no longer makes sense for the fight-or-flight response to always be on.

Not everyone develops post-traumatic stress following a terrifying situation or trauma. But for people whose physiology is susceptible, living in that hyperalert, hyperreactive state for too long can result in medical conditions including (but not limited to) gastrointestinal, cardiovascular, and musculoskeletal complaints, or a variation of one or more of the following symptoms due to what is commonly labeled post-traumatic stress disorder:

- Anxiety
- Rumination or sense of doom
- Hair-trigger anger
- Hypervigilance
- Suicidal ideation
- Poor sleep/nightmares

PTSD has been an accepted diagnosis since 1980. According to the APA, after a traumatic event, "shock and denial are typical. Longer-term reactions include unpredictable emotions, flashbacks, strained relationships, and even physical symptoms like headaches or nausea. While these feelings are normal, some people have difficulty moving on with their lives."[10]

Humanity is in turmoil, with wars, political or resource upheaval, natural devastation, or disease playing out across the globe. Eugene and other authorities have pointed out the long-term trauma consequences of the COVID-19 pandemic, for instance.[11] "People around the world are likely to develop post-traumatic stress as a result of the trauma around the COVID-19 crisis, whether due to the chronic symptoms, the isolation or death of a loved one, abuse at the hands of a parent or partner while under lockdown, or the overwhelm of first responders and hospital workers."

An exact percentage of those with PTSI symptoms may elude statistical analysis for many years to come. According to Eugene, people deny that the clinical definition applies to them, being historically defined by the American Psychiatric

Association as a disorder emerging from something "outside the range of usual human experience, like war, torture, rape, and natural disaster."[12]

Eugene believes that criteria is based on a broad, harmful assumption and that science and biology prove otherwise. "I took care of a kid, kind of a quirky guy with anxiety issues, who was walking down a hallway in his high school when someone came up from behind and touched him on the shoulder. Super sensitive, he jumped. Worse, that pushed him into PTSI—it activated his fight or flight and it stayed activated. Under the DSM-III conventional 'wisdom' (the version of DSM available at the time), he wouldn't have been diagnosed."

Furthermore, there is the continued use of the word *disorder.* The stigma of a mental disorder keeps many (especially combat soldiers) from accepting a diagnosis of PTSD/I. There are also those who suffered long-term micro-stressors who would view their symptoms as something akin to "personality quirks" rather than a disorder. The negative association with mental illness erects obstacles to necessary care and leads to suicides and other avoidable catastrophes, like domestic violence.

Regardless, PTSI patients may find themselves hyperaroused and hypervigilant in situations where other people feel reasonably safe, like walking down a city street. Many suffer sleep disruptions. Some become emotionally numb and withdraw from situations that trigger their symptoms, leaving them isolated. Sometimes, post-traumatic stress causes irrational rage or violence. Sufferers who are startled in their sleep may, for instance, strike or attempt to strangle loved ones. Some people become more depressed and suicidal.

These particular reactions to the stress of trauma have long been recognized, with examples found in historical or literary sources and mythological tales.

According to *Military History Now*, "The ancient Greek historian Herodotus may have been one of the first to write about the emotional strain of combat," telling the story of a soldier who develops hysterical blindness after being nearly slain in a terrible battle between the Greeks and the Persians.[13]

Shakespeare often had soldiers or veterans in his plays reference the psychological toll that battle had on them. In *Henry IV, Part 1*, the character Hotspur is so melancholy and withdrawn after losing friends in battle that his wife is worried, saying, "Thy spirit within thee hath been so at war / And thus has so bestirred thee in thy sleep, / That beads of sweat have stood upon thy brow."[14]

Around the 1600s, *nostalgia* was used to describe despair such as this among the soldiers. During the early 1800s, military doctors began diagnosing soldiers with *exhaustion* following the stress of battle.[15] During the same period but outside the purview of combat, the terms *railway spine* and *railway hysteria* emerged to describe the trauma of surviving a catastrophic railway accident. The symptoms linked with both terms bear a remarkable resemblance to what we now call PTSD/I.[16]

All of this, we are now seeing, is the same—PTSI is driven by the invisible machine.

Following the Civil War, veterans were labeled as having *irritable heart* or *soldier's heart* by Dr. Mendez DaCosta in his 1876 research paper, describing symptoms such as startle responses, hypervigilance, and heart arrhythmias.[17] This was the first biological, not psychiatric, description of PTSI.

Two world wars and the ongoing war on terrorism have given rise to many more colloquial descriptors of post-traumatic stress. In World War I, the terms *combat fatigue* and *shell shock* were introduced; in World War II, *battle fatigue* was coined. Following the Vietnam conflict, *Vietnam syndrome* became popular. Similar terms are found around the world; for example, following the war in Chechnya, Russian war veterans were often said to have *Chechnya syndrome*, highly associated with high suicide rates. "May be replaced by *Ukraine syndrome*. The numbers are going to be bigger," says Eugene.

The wide variety of mental health conditions were codified in 1952 when the *Diagnostic and Statistical Manual of Mental Disorders* (DSM-I) was published. The first edition used the term *stress reaction*.

In the 1980s, *post-traumatic stress disorder* was finally introduced in the DSM-III. Medical classifications today include PTSD and *complex PTSD* (generally assigned to those with extreme symptoms, commonly found in those dealing with high levels of trauma for an extended period of time, like rape victims or combat soldiers or those who come from abusive homes), though PTSD experts have begun to use further classifying terms, including *secondary PTSD* (someone develops PTSD symptoms after sustained time with someone else whose PTSD symptoms are extreme and affect others, for example, caregivers for the terminally ill or the spouses or children of soldiers with PTSD), *comorbid PTSD* (someone has PTSD plus another mental health concern, like addiction), and *uncomplicated PTSD* (linked to one major event, such as surviving a tsunami).

As mentioned, those diagnosed with post-traumatic stress disorder often object to using "disorder" in the name. Jen Satterly, cofounder of the All Secure Foundation and coauthor of *Arsenal of Hope: Tactics for Taking on PTSD, Together*, says, "Our soldiers hate the term PTSD, in particular the word *disorder*. They don't want to be labeled anything that makes them sound sick, like they can't go back into the field, or weak, like they couldn't kick the enemy's ass when they need to. Because of that, some medical or support agencies are now referring to PTSD as PTS or as something else entirely, like occupational stress injury."[18]

And, medically, they are correct to argue against the term *disorder*, according to Eugene. Clinical research offers proof of a measurable, physical change.

THE DINNER CONCLUDES

Jamie had been listening intently to Eugene for almost two hours. The dinner conversation had been far from boring.

"Dr. Frank Ochberg was the first to call it PTSI,[19] and I concur with his findings," said Eugene. "The invisible machine overactivates; it's a biological event. And I can reset it."

Jamie grinned at him. "And for that, I am grateful."

"When something bad happens," the doctor continued, "like the tiger chasing the caveman up a tree, the body registers the trauma. The biological response has been demonstrated for years. There is no debate anymore—this happens." Physical changes to a brain struggling with PTSI are observable with scans. Eugene talked about Reservist Jason Brown's convincing medical records and the longevity of the soldier's relief from symptoms. "I know it sounds like fiction! People will say, no way a simple shot can do all that!

"A few months after the first treatment, Jason was retraumatized by fireworks and underwent a second treatment, but he has reported no reoccurrence of symptoms to this day. The story of Jason is that of a man who returned from a yearlong tour in hell with a spinal injury, brain injury, and complex PTSI. He came in for my procedure and he has been doing well for *fifteen years now*!"

Jamie's last doubts faded. He raised his glass in a toast. "Cheers."

The pain physician smiled. "By the way, my innovation was recommended by President Obama in 2007, and it is my understanding over a thousand of these procedures are being done a year among Special Forces at Fort Bragg alone. So, you know, boom. Mic drop. This is real."

As far as Jamie could tell—based on preliminary research and his own successful experience—Eugene really had found a way to reverse the effects of post-traumatic stress far more effectively than any other approach out there. It had taken years to get him to this point, but he had not given up.

"Even knowing there was a physical reaction happening between the sympathetic nervous system and the amygdala, it took me years to figure out *why* an injection of local anesthetic next to ganglion nerves in the neck worked on chronic traumatic stress and why then it calmed down an overactive amygdala in the brain. Understanding this became very important to me."

As the night progressed, Eugene shared his personal story with Jamie. It became clear why the man had made a career out of relieving people of misery due to trauma.

CHAPTER TWO

Resetting the Machine

Eugene Lipov's childhood was laced with terrible events, forced moves, clinically depressed parents, and sporadic violence in the home. He was writing suicide notes by the time he was eight.

Born in Ukraine, Eugene's father was a highly qualified surgeon forced to work in a TB clinic with an infinite line of diseased patients hacking up blood. Their Jewish family lived in poverty, treated poorly by the community. When Eugene was six, he and his friends found an unexploded land mine while playing in the local woods; Eugene told his father, who then forbid him to reenter the forest. His friends were not restrained and went back only to have their limbs blown off when the mine exploded. In this antisemitic environment, his father was unfairly blamed for the maiming and deaths of the children.

Eventually, his father was forced out of the country. Eugene, his brother, and his mother (also a doctor) followed him to the United States a year later, in 1973, with few possessions and no language but Russian.

As a nonathletic, awkward boy with few friends, Eugene struggled to assimilate. Later, he found his footing in academics, receiving a bachelor's degree in biochemistry and then a medical degree at the prestigious Northwestern University. However, he remained dogged by his family's bad luck and depression. During his surgical residency, Eugene was in the middle of an operation, his gloves and surgical gown covered in blood, when he received an emergency call—his mother had killed herself. This despite her prescribed pharmaceuticals and regular visits to a psychiatrist.

Deeply shaken and unable to focus, he felt compelled to leave the surgical residency, and he moved into the field of complex anesthetics. He spent two years in an anesthesiology residency at the University of Illinois before completing his training at Rush-St. Luke's Medical Center. There, he underwent advanced training in pain management with a focus on pain intervention. He'd become interested in how pain, mental and physical, damaged the body—and what he could do to stop it.

Now an anesthesiologist, Eugene discovered that the widely used SGB procedure not only relieved headaches and nerve pain but, with a slight adaptation, also successfully relieved the discomfort of severe hot flashes in postmenopausal women.

"My older brother, Sergei, is an internist, and we were discussing my patient, who had terrible postmenopausal hot flashes and nothing was helping," Eugene would later explain. "As we talked, we realized that hot flashes were similar to complex regional pain syndrome, a type of burning arm pain that the SGB is used to treat. I mean, hot flashes are burning discomfort everywhere."

With that realization, Eugene did an SGB treatment on the patient, Christine, who had such bad hot flashes that she did not wear a coat, not even during the cold Chicago winters. The injection gave her immediate relief—she went eight weeks without having another hot flash.

That led him to treat a number of other women for hot flashes with the SGB. A published study in 2008 reflected Eugene's success with thirteen survivors of breast cancer in remission who suffered severe hot flashes and night awakenings, finding that the stellate ganglion block relieved most of the thirteen patients of the flashes and sleep deprivation with no side effects.[20] This was published by *Lancet Oncology*, the number two medical journal in the world.

Since then, medical studies have shown that relief from hot flashes wasn't a coincidence.

But really the challenges came because he couldn't explain exactly *how* his innovation worked on postmenopausal women yet. The *Chicago Tribune* even did a hit piece, describing the procedure as a "long hypodermic needle plunged into [the] neck" of the patient Bianca Kennedy, with one expert quoted as saying SGB's benefits for menopausal women were a "crock" and that they should instead try yoga and drinking cold water.[21]

Yet, despite the inflammatory piece, the writers acknowledged that the efficacy of the procedure was real—with no side effects. "Kennedy says it is the only remedy that has ever worked for her," they wrote.

Because it *did* work. On hot flashes and, Eugene would soon find, more. Much more.

YES, BUT WHY?

The original block, the SGB, was a well-established pain management procedure that originated back in the 1920s, when some wise person somehow discovered that injecting a local anesthetic next to the sympathetic ganglion nerves in the neck interrupted the pain signal the nerves sent to the brain and relieved migraines and chronic pain. Dr. Lipov first pioneered the modern-day use of SGB to eliminate hot flashes among postmenopausal women in 2004, and Lipov's Advanced Pain Centers began treating hot flash patients in 2005.

A few years later, when the data and studies came in about his success with severe hot flashes,[22] Eugene was recognized around the world as an innovator. He had developed a highly successful private practice by now, and, despite early detractors, the Norwegian government brought him over twice to train their top health officials on how to use his innovation.

But it bothered him that he still struggled to explain exactly *why* this treatment was working the way it did. He spent his free time revisiting all the research published on the stellate ganglion block, as well as researching medicine and biology outside his field from every specialization, consuming more than three thousand scientific articles. He knew the intersections were there; he just had to find them.

It was then he discovered a report by a Finnish doctor that dealt with hand sweats but was fascinating in an unexpected way. During surgery, the surgeons had clipped the T2 ganglion clumping of nerves around the spine in the chest on a patient with severe hand sweats. They found that not only did the sweats go away, but so did the patient's anxiety and other symptoms of PTSI.

In a conversation about this with his brother, Eugene said, "Whoa, that is crazy. Why should that be?" But then it occurred to him that, neurobiologically, the part of the sympathetic nervous system called the *thoracic ganglion* is very similar to the stellate ganglion. Anatomically, the T2 ganglion block used in the chest is the same thing as the stellate ganglion block.

Eugene started paying attention to other outcomes for his menopausal patients and found that some were reporting alleviation of what would be considered post-traumatic stress symptoms.

He called his brother again, who became very excited and agreed that the pain physician should start applying the procedure to those with post-traumatic stress. He already knew the century-old procedure was safe, with minimal side effects.

Eugene eventually discovered he could reconfigure the SGB procedure and perform what essentially was a sympathetic nervous system "reset." His first PTSI patient, sent by his brother, was on the threshold of being admitted to a psych ward because of extreme and unmanageable symptoms. The patient underwent the procedure and woke up on Eugene's clinic table a different man—the man he used to be before the trauma. The patient felt relief from his worst symptoms within thirty minutes.

Somehow, applying the anesthetic next to the stellate ganglion in the neck reversed the biological changes brought on by trauma, bringing the sympathetic nerve back to its pre-trauma state. Again, Eugene couldn't then explain the minute details of why there was success. But he would.

PROVING HIMSELF

"You are so full of bullshit," Robbin Lipov said to her husband one morning at the breakfast table, as he was going on about his new findings. "Nothing can cure everything."

She was teasing her husband. But Eugene knew there was truth to what she was saying. It seemed too good to be true. He had to prove to her that he could continue to transform SGB and inhibit post-traumatic stress symptoms. He had to prove it to himself, too. So, he set about writing a peer-reviewed paper with his brother and six other physicians, sharing his findings with the world. And, of course, his wife.[23]

After that article was published, Dr. Lipov presented a paper in New York regarding ten patients who'd had their trauma symptoms alleviated within thirty minutes of the procedure. Medical experts started reaching out to him more frequently, including one who suggested Eugene seek out Dr. Stephen Porges, a respected physiologist who focused on the parasympathetic nervous system in Chicago.

The sympathetic nervous system (controlling fight-or-flight responses) and the parasympathetic nervous system (responsible for "rest and digest" activities) are both parts of the autonomic nervous system. Porges was a leading figure in neuroscience and had introduced the groundbreaking polyvagal theory. "He is considered a 'god' by the yoga people," Eugene will say with a smile.

Though Eugene continued to hold a deep distrust for those related to the psychiatric field after watching how his mother's care ended in her suicide, he arranged a short meeting with Porges out of curiosity. The two men ended up in a discussion for more than five hours, developing mutual admiration for the other's line of research. "I felt like we were bonded like brothers by the time I left," Eugene said.

Porges also introduced Eugene to the world-renowned psychiatrist Dr. Frank Ochberg, who, along with Dr. Jonathan Shay, first defined the term *Stockholm syndrome* and had a career in research and application revered by his peers. Interestingly, Dr. Ochberg had cowritten a book decades before, titled *Violence and the Struggle for Existence*, published in 1970 (Little Brown)—two years after Martin Luther King Jr.'s assassination, with a foreword by Coretta Scott King. A chapter called "Biological Bases of Aggression" theorized that new, unwarranted aggression in an individual came from biological impulses—a theory proved true when Eugene was able to identify and eventually explain the process of the mechanism that initiated these biological impulses.

Between Porges and Ochberg, Eugene suddenly had access to a host of brilliant psychologists and psychiatrists who allowed him to revamp how he thought about their field and use their research to apply to his own work. Eugene came to view trauma therapist Dr. Peter Levine's work on an alternative therapy called *somatic experiencing*, which was aimed at modifying the trauma-related stress response, as essential and a huge part of trauma recovery. Somatic experience was akin to "installing new software" for the patient, as it were, once their "invisible machine" has rebooted.

In a 2012 interview, Dr. Ochberg argued publicly that the label *PTSD* should be changed to *PTSI*. He stated, "There is a crisis of suicide, stigma, and misunderstanding affecting young veterans . . . the name affects civilian survivors of trauma, as well . . . the concept of an injury, rather than a disorder, does justice to their experience. Once they were whole. Then they were shattered."[24]

Eugene agreed with Ochberg's premise and the name change, but took it further. He believed there truly was a physical mechanism injured by trauma. A disorder of the brain would have inconsistent symptoms, but trauma patients had the same basic symptoms, regardless of whether the patient was a combat soldier, a sexual assault or child abuse victim, a first responder, or anyone who dealt with a slow drip of toxic stress over time. For the primary symptoms to be the same (differing only in degree), no matter the type of trauma, it made sense that *an actual biological change* to the sympathetic nervous system was at play.

There was also the fact that a simple injection of local anesthetic applied to the ganglion nerves reversed the symptoms, no matter the traumatic experience.

THE REFINEMENT

In 2014, a patient with back pain came in to see Dr. Lipov. The man had been a sniper with the Marines. One day, the patient returned to the clinic and told Eugene he was going to kill himself, but he wouldn't let the pain physician admit him to a hospital. He knew Eugene had been working with trauma patients.

"His wife was crying, he was crying," Eugene recalled. "I told him he was putting me in a bad position and he told me 'tough shit,' that it was about him and not me, and he would kill himself if I made him go to the VA. He wanted me to give him the SGB injection."

Eugene performed the block at the C6 vertebrae, per usual. Forty minutes after the injection, the ex-Marine said he was still going to kill himself, that he felt no difference. At that point, Eugene was frightened; he wouldn't let the patient leave. He decided to try the block at the C3, something he had tried while working with hot flashes in Norway. He performed the second injection and waited with bated breath. Luckily, just five minutes after the procedure, the patient smiled and reported that his overwhelming doom was gone. He felt great, he said, and didn't return with symptoms.

"It worked," Eugene said. "It turns out, when I was doing only the C6 all those years, I just wasn't doing enough block. When I do C3/C6, I have a bigger area of nerves being reset."

Thus, the dual sympathetic reset innovation was born.

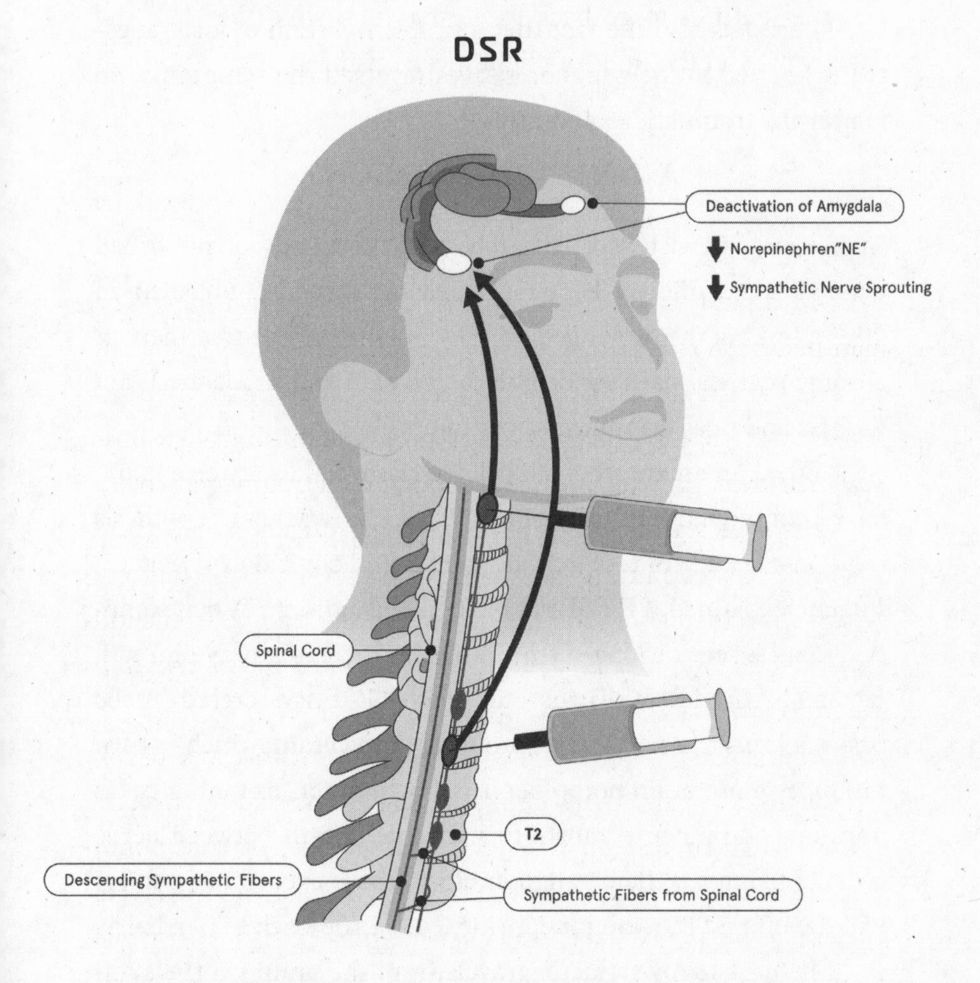

Dual sympathetic reset uses two injection sites, at C3 and C6.

It still took Dr. Lipov a few years to get a clear picture of what happens when a trauma puts this invisible machine into overdrive and then refuses to go back to normal, and what the DSR did exactly to reset the fight-or-flight mechanism.

A UNIFYING THEORY

In the end, persistence, research, and thousands of published clinical trials offered Lipov a unifying theory (published in 2009), linking the prolonged efficacy of DSR for the treatment of chronic regional pain syndrome (nerve pain and headaches), hot flashes, and post-traumatic stress injury.[25]

"Why can an anesthetic block that lasts eight to twelve hours have a consequence of years of relief? All I can offer is a hypothesis built on a decade of research, diligence, and clinical experience," Eugene explained. "PTSI is a biological condition. When somebody has severe or long-term trauma, it increases nerve growth factor (NGF) levels. Turns out, when NGF is secreted in the brain, it leads to nerve fibers sprouting in the brain, which in turn produces more brain norepinephrine (which acts as both a stress hormone and a neurotransmitter sending signals between nerve cells). During a stellate ganglion block, those excess nerve fibers will die off and the norepinephrine levels drop back to normal."

The nerve growth factor travels from the brain via the sympathetic nerves leading to the stellate ganglia. The bigger the trauma, the bigger the NGF response. That nerve growth factor then stimulated new nerve "sprouts" of sympathetic fibers in the brain. If the trauma was prolonged, NGF continues to be sent to the stellate ganglia, and more nerve fibers are generated, producing higher levels of the norepinephrine. These extra nerve sprouts are believed to be keeping the system stuck in "fight or flight," pumping norepinephrine; increased norepinephrine levels overwhelm the system and signal to the brain that the stress or danger is ongoing.

"If you were to sample the cerebral spinal fluid around the brain in a soldier with PTSI, you would see the norepinephrine level twice as high as normal," said Eugene. Where once there were four nerve fibers, there may now be ten new nerve sprouts, which means an alarming amount of norepinephrine coursing through the amygdala and the rest of the brain.

"This is the cause of post-traumatic stress," Eugene said. "So, what am I doing to reverse the symptoms? To reset the machine, as it were? SGB/DSR has been shown to lead to pruning of extra sympathetic nerve fibers, likely bringing the fight-or-flight mechanism back to the pre-trauma state."

He applied local anesthetic next to the ganglia nerves to reduce the nerve growth factor. The new nerves died off. The procedure resets the sympathetic nervous system, and the amygdala normalizes. Often, if additional treatments are needed, any regression becomes less and the treatment becomes permanent as the excess sprouts are "pruned" back completely to the pre-trauma state.

Technically, three sympathetic ganglia exist in the neck: stellate ganglion (SG: C7–T1), middle cervical ganglion (MCG: C6), and superior cervical ganglion (SCG: C3). SG and MCG supply the part of the brain via one artery, called the *vertebral artery*, while SCG supply the part of the brain via one artery called the *interval carotid artery*. A combination of SGB and SCG produces a more complete reset of the nervous system. The overactivity in the brain returns to normal.

"The two injections are done on the right side of the neck near the C6 and C3/C4 vertebrae, and then two more injections on the left side, if necessary," said Eugene. "We use an ultrasound for exact placement. The key here is that the sympathetic nervous system has a feedback loop that continues to activate the amygdala. SGB, now DSR, interrupts this feedback loop."

According to Eugene, "Once you anesthetize the stellate ganglia nerves with the local anesthetic, the norepinephrine levels drop in five to ten minutes. The overactivation is reversed that quickly. That is pretty damn amazing. And it is why people feel markedly better very quickly."

With the refined dual sympathetic reset procedure, 80 to 85 percent or more of the PTSI symptoms disappear within hours in the majority of patients, with minimum side effects. If the patient has exceptionally severe PTSI, or underlying health issues that interfere with the injection placement, or is re-traumatized, they may require an additional booster.

The dual sympathetic reset basically turns down the biological noise. However, the trauma endured by a patient is still emotionally impactful, especially if it was, as Eugene says, "Big F-ing Trauma. BFT." This procedure gives almost anyone a genuine shot at living in a joyful world again, where they can sleep and function at a healthy level, but it also allows them to do the work required to heal emotionally. Cultural or environmental problems may continue to be an issue for an individual; these problems definitely continue to be sources of deep distress on a societal level. The DSR can reset the nervous system and alleviate the symptoms, but the systemic causes may still be in play, which will need to be acknowledged by patients seeking to avoid having the fight-or-flight response retriggered.

For those who also suffer from traumatic brain injury (TBI), the best outcomes are accompanied by medical intervention for both the TBI and the PTSI. Obviously, any talk therapy or useful unpacking of one's trauma will be easier and more effective if a biological issue causing extreme mental discomfort is gone. A physical therapist would not do physical therapy without first resetting a broken leg.

THE MEDICAL COMMUNITY STARTS TO TAKE NOTICE

During a flight in 2012, Eugene called a friend from his first-class seat on the way to speak at the Pan-Asian Oncology meeting. "This world is so crazy! I come from this poor place in Ukraine, where my father's patients are coughing their blood all over me, to being flown first class around the world. Crazy!"

In 2015, Dr. Michael T. Alkire and colleagues presented a report to the American Society of Anesthesiologists: "We found SGB had efficacy for significantly reducing PTSD symptoms in a rapid and sustained manner that allowed functional brain activity to be compared in the same subjects when they were suffering with PTSD symptoms versus when their symptoms were greatly diminished."[26] In this important paper, Alkire addressed the clear evidence of clinical improvement *and* the use of scans before and after the procedure reveal that the amygdala's overreactivity was reduced or eliminated.

Finally, the medical community was acknowledging that Eugene had found a true game-changer in physically reversing the overactivated sympathetic nervous system, a biological issue once considered a mental disorder. He could reset the invisible machine and take the chronic pain of trauma away.

And by 2019, he was no longer alone in his efforts. That was the year Steven Taslitz, chairman and cofounder of the multi-billion-dollar private equity firm Sterling Partners, met Eugene. Steven had learned of him through a patient, a grateful Marine who had found relief from his PTSI. Since that meeting, Sterling Partners has been working with Eugene to open Stella clinics around the United States (and the world), bringing DSR to those in need.

THE REALITY OF "RELIEF" HITS HOME

For Eugene, the successes he was seeing meant that there is help for people in misery because of past trauma. Help for people who have suffered through religious persecution and war atrocities, as his family had back in Ukraine. People like his mother, who had taken her own life after years of mental anguish.

Help for people like himself.

A few years ago, a stressful event triggered and heightened the post-traumatic stress symptoms Eugene had struggled with due to his traumatic childhood and his mother's sudden death. He would come home and find himself immobilized.

"You can function at work fine, but you come home and you are a limp noodle. I need help with our son," his wife repeatedly told him.

Finally, Eugene reached out to his friend Dr. Porges, who said, "You're having fugues. You're going into a fugue state, associated with trauma dissociation."

According to WebMD, "A fugue in progress often is difficult for others to recognize because the person's outward behavior appears normal. Symptoms of dissociative fugue might include the following: Inability to recall past events or important information from the person's life; Confusion or loss of memory about their identity, possibly assuming a new identity to make up for the loss; Extreme distress and problems with daily function."[27]

Realizing the truth of what his friend was telling him, Eugene joked, "Oh, hey, thanks for the input. You couldn't have told me earlier?" After an official diagnosis, the doctor became the patient. And, as he'd seen with so many people over the years, Eugene's relief was immediate. His blood pressure went down. His cholesterol went down. He could sleep again. He was "emotionally happier, which made my wife and son happier." And his fugues went away.

This only reinforced what he knew to be true:

Trauma no longer needs to ruin lives.

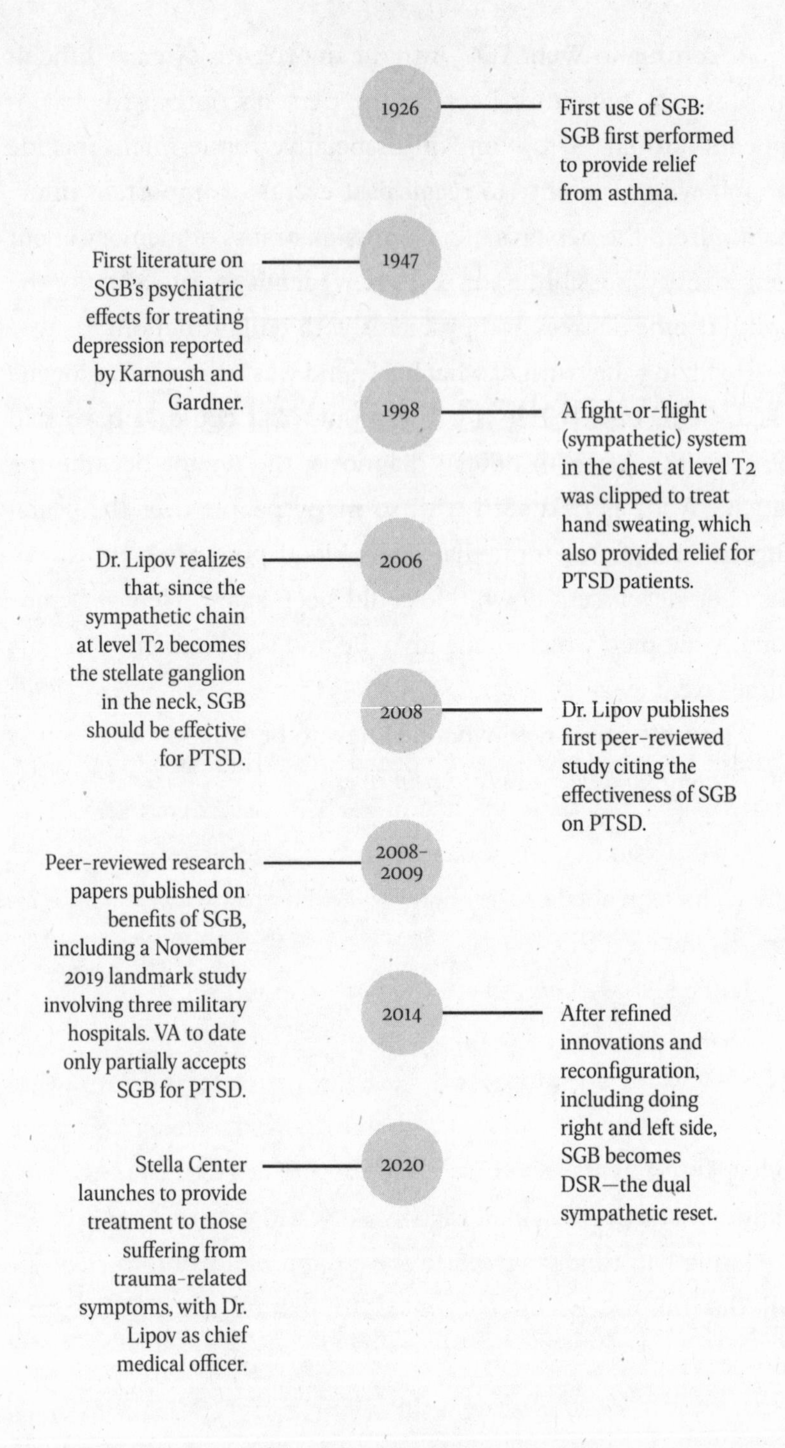
1926
First use of SGB: SGB first performed to provide relief from asthma.
1947
First literature on SGB's psychiatric effects for treating depression reported by Karnoush and Gardner.
1998
A fight-or-flight (sympathetic) system in the chest at level T2 was clipped to treat hand sweating, which also provided relief for PTSD patients.
2006
Dr. Lipov realizes that, since the sympathetic chain at level T2 becomes the stellate ganglion in the neck, SGB should be effective for PTSD.
2008
Dr. Lipov publishes first peer-reviewed study citing the effectiveness of SGB on PTSD.
2008-2009
Peer-reviewed research papers published on benefits of SGB, including a November 2019 landmark study involving three military hospitals. VA to date only partially accepts SGB for PTSD.
2014
After refined innovations and reconfiguration, including doing right and left side, SGB becomes DSR—the dual sympathetic reset.
2020
Stella Center launches to provide treatment to those suffering from trauma-related symptoms, with Dr. Lipov as chief medical officer.

CHAPTER THREE

Stumbling Out of the Dark

YES, YOU

Months before Jamie had even heard of Dr. Eugene Lipov, he'd surprised himself by making an appointment with a counselor.

Six weeks into their powerful conversations, the joints on the old office chair creaked as the therapist asked vehemently, "How could you *not* have PTSD?"

Jamie stared at her, unsure how to respond.

"Can't you hear yourself? Seriously, how could you not have PTSD?" The question was offered with wet eyes. An androgynously dressed pixie with short, sharply parted hair, tailored cardigan, fitted pants, and button-down shirt, this person had always maintained a professional, slightly detached demeanor until now.

Jamie had finally agreed to see a counselor after years of internal discomfort, which continued to mount steadily despite his growing reputation and clientele base. His most recent relationship had dissolved; he had been milling around the house alone for months, with more than ample opportunity to evaluate his mental health. However, he hadn't expected this session to get emotional—especially not by the therapist.

The question shifted something inside of him, uncoiled deep inside his mind, dropped into place, yet . . .

Post-traumatic stress disorder? Come on, he thought.

Jamie was a middle-aged man born to a white father with Asperger's and a black mother, two apparently brilliant people who were MIA while he grew up in the darkest part of sunny LA, but he believed his story wasn't any different than millions of other kids with shitty childhoods. He hadn't suffered years of sexual torture. He hadn't served in the military. He wasn't a firefighter. He hadn't survived a natural disaster or a war.

He let the therapist hug him. This person's short arm span made him feel like a giant, but the spontaneous show of caring comforted him. He felt seen. Validated. Acknowledged as a human in real distress, walking alongside all the other humans. Eventually, Jamie dropped back into his armchair.

"Listen, I am not a soldier dodging bullets all day." He blew out a noisy breath. The small office filled with his frustration. "I didn't watch my buddies get blown up by a car bomb. I didn't pull screaming, dismembered people out of the rubble on 9/11. I don't have PTSD."

The therapist seared him with a gaze. "You are such an accomplished guy, Jamie. You are a branding authority, an artist, and a noted speaker and author. You have a degree from that school in London. You run in important circles, travel the world, and live in a nice house. You're a hard worker. You are normal.

That's what you are saying to yourself, right? That you don't have mental issues?"

"Exactly," he said. "Feeling isolated or anxious doesn't mean I have a mental disorder. Maybe that's how an artist is supposed to feel."

"I don't think you or anyone else with PTSD is mentally unhinged." The therapist tapped a stack of notes on the desk. "But you can't sleep at night. You are uncomfortable when people get too close. You can't celebrate when your hard work pays off, only feeling more discomfort and unhappiness. I could go on."

Jamie waved a hand in front of his face, trying to swat away the words. "But—"

"Jamie, you've been telling me about growing up in slums in Los Angeles, abandoned as a baby, deprived of food, housing, safety, and most importantly human touch, too white to be black and too black to be white, stuck with abusive caretakers or thrown away, barely literate until your late teens. If you were to hear this story from someone else, would you not agree that they had spent their childhood in a state of completely overwhelming stress and trauma?"

HUGGING THE CACTUS

Jamie was strangling. Being strangled.

He jolted awake, yelling—though, thankfully, it came out as a hard mumble behind closed lips. The artist threw back the blankets, frustrated, exhausted, and crawled out of bed.

A few months after that conversation with the therapist, the Oregon summer faded into a wet fall. As Jamie sat waiting for the coffee to finish brewing, wide rivulets of rain on the windows blurred the fir trees beside the house and the Portland skyline beyond.

He couldn't stop thinking about the idea of post-traumatic stress.

He had argued with the diagnosis, but his instinct told him it was right on the money. *Nightmares? Check. Hypervigilance, edgy no matter where I am? Check. Severe anxiety and social isolation? Check. Irritable and self-destructive? Check. Loss of pleasure or interest in favorite activities? Check and double check.*

He'd thought these were the attributes of an artist-introvert, but the therapist had nailed him to the wall. He had to stop denying the truth of who he was.

The loss of pleasure had been the final straw for him. He'd landed some of the most significant contracts of his life, a project with Adidas and then a talk at Nike. Yet, all he could think was, *This is my dream, to be working with these brands and making their appeal more emotional, more connective. So, what's wrong with me?* Because he felt . . . nothing. Except for a growing unease that would seep up whenever he took a second to stop and breathe. Why? Jamie couldn't understand what was happening to him. He'd signed his first book deal, and it was the worst week of his life. He drank champagne with his friends, grinning at the celebratory backslaps, but it was all false, his mind swaddled in that ever-present doom, sure the other shoe was about to drop.

What shoe?! Why think that?

His dreams were coming true, and he couldn't enjoy them. He had to do something.

Jamie had an aversion to doctors and definitely anyone in the psychiatry field, having grown up very poor and under the delusion that spending money on personal health, especially mental, was an unnecessary waste of resources. But a friend had talked him into seeing a qualified mental health counselor, because one, he was being an asshole, and two, as she said, he could damn well afford it.

After the Portland therapist, he had reached out to a trauma specialist he knew (they shared a literary agent): Dr. Shauna Springer focused on the realm of post-traumatic stress. She was a renowned military trauma doctor who had been at TAPS (Tragedy Assistance Program for Survivors) for years but worked out of the Stella Center in Chicago. And she had told him about a procedure her boss had innovated, a procedure that could get rid of the symptoms of PTSD.

However, instead of saying *PTSD*, she referred to it as *PTSI*: post-traumatic stress injury. A physical overactivation of the sympathetic nervous system to be healed, like a broken bone or a bacterial infection. A term that made Jamie's wound sound real, not so ethereal.

The doctor helped him visualize the sympathetic nervous system as a railway station of nerves in the neck called the *stellate ganglion*, situated on either side of the voice box. She said all it would take was an injection or two in the neck and the nervous system would be reset; it would go back to a pre-trauma state.

It sounds like goddam science fiction, Jamie thought. But even if the procedure was somehow genuine and effective and didn't turn patients into zombies, he struggled to believe himself worthy of that kind of medical treatment. Soldiers and first responders were the ones who deserved intervention, not him.

Finally, if a cure was real—and that was a big if, considering no one seemed to know about this miracle procedure except a handful of doctors and researchers—what happened if he had the injections and then realized he no longer was able to produce art? Wasn't his angst and stress a big part of what drove his creativity?

But all the thorny shit he'd been trying to avoid for years was front and center now, and it was not helping his work or his relationships. Thanks to the counselor, Jamie had finally faced the fact that he'd been severely victimized and manipulated throughout his childhood. That it wasn't normal to look back on living in the falling-down Mexican-Nicaraguan-Armenian LA neighborhood with eight people in a one-hundred-square-foot apartment as the "good" times just because he wasn't in an institutional environment. That Jamie was a victim, no matter how much that word offended his sense of self. Facing these truths in himself, while the world outside his windows was a lush, thriving green . . . his equilibrium had unraveled.

Jamie was ready to hug the cactus, he told himself. Grab on to the pain, embrace it, and then squeeze it out of himself. He was terrified of the possible consequences of this procedure, this injection, but he was more afraid to do nothing. *How much will I lose if I just keep on keepin' on?*

He certainly could never have foreseen how his next phone call that rainy afternoon would establish a chain reaction that would change his life—and bring together a small group of random strangers who might just possibly change the world.

A PROFESSIONAL OPINION

Dr. John "Jay" Faber is a double-board-certified trauma physician who works at the Amen Clinic in Encino, California, a nationally known treatment center.

But Jamie hadn't known any of that when he first met Faber. Their introduction had happened randomly. Faber had read Jamie's book *The Iconist* and emailed to ask him to speak to a group of troubled inner-city kids in LA for his charity (the Faber Ryan Youth Foundation) about redefining themselves and making themselves stand out in a positive way in a noisy world. Jamie had jumped at the chance to give back to a community of youth he knew well.

Faber had written a well-received book on criminal justice and brain injury;[28] he is one of the leading forensic psychiatrists and a regular expert witness in criminal trials throughout the US. In addition, Dr. Amen has successfully worked to restore damaged brains and has published articles on his interventions based around multiple modalities, which included rehab, pharmaceutical agents, and supplements. Amen had also become a nationally recognized trauma physician, utilizing brain SPECT (single photon emission computed tomography) scan technology that helps assess malfunctioning distinct neuroanatomic regions leading to holistic treatments that avoided polypharmaceutical drug use whenever possible, proven to help reliably restore brain function.

Reading Faber's book, *Escape: Rehab Your Brain to Stay Out of the Legal System*, was what got Jamie first thinking about brain health in general and his own in particular.

Now, Jamie had questions, and Faber was a guy very much in the know. The artist felt comfortable reaching out and asking what the doctor thought about what was happening at the Stella Center—if he thought this procedure was real and safe.

When Jamie called, Faber answered on the first ring.

"A stellate ganglion block? Yeah, I've heard of it," said Faber. "But what does that have to do with you? Isn't that procedure used to stop hot flashes?"

The Lipov procedure, Jamie knew, was based on SGB, a proven, safe methodology and medicine that had been around for a hundred years. DSR was supposedly doing something new and miraculous for people with post-traumatic stress.

Was it too good to be true?

"Here," Jamie said. "I'm going to send you a couple of links. It sounds like this Dr. Lipov might be the real deal. But I'm not going to move on this if you tell me it's ridiculous."

"I'll look into it, but the decision is yours, Jamie. It boils down to three questions: How invasive is it? What are the side effects? Is the potential upside worth the risk? Let me review some of the literature, and I'll send you some articles. After you've evaluated the answers, and if it is noninvasive with little side effects and high efficacy, then you could consider going for it."

TO GO OR NOT TO GO

Faber's risk-versus-reward evaluation was heavy on the reward. The trauma physician had done a deep-dive analysis, curious about something that could benefit his practice. He found that the procedure was noninvasive and moderately affordable since it was an outpatient procedure using the same two-dollar amount of anesthesia utilized in an epidural. There were no side effects. There was no downside.

The upside was that since 2008, numerous reports documented how what is now called the *dual sympathetic reset* had rapid effects on reducing PTSI severity among veterans, active duty service members, and civilian populations. According to Faber, Eugene's innovations were significant. However, the fact that the basic procedure was rote and routine and a century old did communicate its effective history in treating pain.

There were stray naysayers, but the qualified research and documented success outweighed their objections (discussed in detail in chapter ten).

Jamie felt he should have been convinced by Faber's three qualifications and his final thumbs-up. The decision seemed like a no-brainer . . . and, yet, his brain did require a bit more. He still had the one concern, and it was a biggie: *If this DSR really works, what will happen to my artistic drive? Stress has been my prime motivator for years.* Jamie didn't want to lose what he considered his core identity.

His contact at the Stella Center, Dr. Springer, answered that question: DSR does not remove motivation or drive. Artists retain their creativity. Soldiers retain their reflexes. Athletes retain their response time. Oftentimes, creativity, reflexes, and response time are improved.

For example, elite athletes came to the center for the DSR procedure after accidents or traumatic events hampered their performance. These athletes reported that their reaction time went back toward normal, if not improved, and their competitive nature remained intact.

One case study from Womack Medical Center at Fort Bragg showed that 166 active duty soldiers were treated with DSR and had no reduction in reaction time. Womack Hospital was performing thousands of these procedures, after issuing a report that concluded, "Selective blockade of the right cervical sympathetic chain at the C6 level is a safe and minimally invasive procedure that may provide durable relief from anxiety symptoms associated with PTSD."[29]

Well, he thought, *if elite athletes are doing this and becoming more elite . . . that's cool.*

If athletes and soldiers weren't losing their edge, who was Jamie to worry? This information helped him not only to feel secure but also to view PTSI through a different lens: he didn't see elite athletes traumatized by their injuries or experiences as "broken people," so choosing the same medical intervention did not mean he was giving in to being a broken person.

In the middle of COVID hell, Jamie booked a flight to Chicago and a hotel with an outdoor, rooftop bar that was still seating during the pandemic. The plane was nearly empty; he had rows to himself. His mind's eye kept replaying scenes from *Vanilla Sky,* an old Tom Cruise sci-fi flick.

This is some high-level medical-tech shit, set in a dystopian reality, Jamie thought. *Resetting these nerves that are lying to my brain? Yeah, okay.*

JAMIE GETS HIS JOY BACK

The gloom of a wintery Chicago morning waited outside as Jamie lay on a surgical table in the Stella Center, squinting at the tiled ceiling in a room painted warm, reassuring colors. Dr. Lipov was performing the procedure, having come in beforehand to introduce himself and walk his patient through the ten- to fifteen-minute process.

Jamie had opted to have the procedure on only one side of his neck. He'd also opted to forego the twilight anesthesia and stay awake for no discernable reason other than ego. Two injections were made on the right side, offering a slight pinch at the C3 and C6 vertebrae, about halfway down the neck.

Eugene ensured correct placement of the needle with the use of an ultrasound. It was the same local anesthesia being injected into Jamie's neck that was used in epidurals associated with childbirth, as was the ultrasound machine. For some reason, that pleased Jamie.

"Done," the doctor said, his eyes crinkling in a smile. "How do you feel?"

Uncomfortable but not in pain, Jamie had kept his eyes shut most of the time. When it was over, the right side of his face felt a slight tingling sensation, one eye was watery and slightly drooping, and his voice was gravelly; Eugene had told him to expect this; the drooping and so on is in response to the nerves being "numbed" and is called *Horner's syndrome effect.* According to the pain physician, SGB practitioners knew the injection had been placed improperly if the Horner effect *wasn't* in evidence. This is also true of Dr. Lipov's updated version, DSR. It would go away over the next few hours, Eugene assured him.

Jamie mumbled, "I'm a little tired, Dr. Lipov." In truth, he felt mostly a strange, unfamiliar calm.

"You might feel tired right now, but you're going to feel great in about seven or eight hours." Eugene put a hand on his shoulder. "My colleague Dr. Springer told me to treat you like a VIP. Why?"

Distracted, a cascade of emotions trilling through him, Jamie sat up and rubbed at a phantom pain at the injection site.

"Uh, maybe because I'm an author. I guess? And I know a lot of extraordinary people. I've got a platform . . ." He petered off. *I sound like a self-aggrandizing moron.* All he wanted to do was take a minute to collect himself.

"Hmm. Well, Dr. Springer was adamant I take care of you. What do you think about a late dinner? And you can call me Eugene."

And so began the surprising, powerful relationship that would set an even more surprising row of dominoes into a topple—the chain reaction Jamie had unconsciously birthed with his first call to Faber about DSR.

RIGHT VERSUS LEFT

Jamie walked out of the procedure room with a bit of a sniffle on the right side. It soon disappeared and, within an hour, Jamie realized he did perceive change: calm and clarity.

It's like a curtain of cobwebs has been swept away, he realized. Sounds were more pleasant, complex. And, clichéd or not, colors were sharper, deeper. He and a friend went into the Art Institute of Chicago that day, and it was one of the most visually impactful experiences of his life.

He certainly was not the first to report enhanced sensory perception, especially with colors. In February 2021, Dr. Springer spoke with *Forbes* magazine about this very thing: "Many of my patients say just after the injection that 'the colors in the room look much brighter than before' or that they 'can see more of what's around them with more clarity.' My working theory is that since PTS causes 'tunnel vision,' which I've learned has a medical name—*foveal vision*—that when those who suffer from trauma get calm in their own bodies again, thanks to stellate ganglion block, their tunnel vision literally expands and they are able to take in a greater, more vivid visual array."[30]

Jamie and his friend then went to the swanky restaurant for the aforementioned dinner with Eugene. "I'm not sure I can put into words the many epiphanies and 'what ifs' that occurred to me that night, both from my amazing reaction to the injection and to my conversation with the smartest human I have ever met, at a dinner that felt like it had to have been arranged by the gods," Jamie later recalled.

For three hours, they talked nonstop and drank expensive red wine, paying little attention to the pouring rain and occasional siren just outside the open windows. The high-end Mexican food was not nearly as good as the mom-and-pop taco trucks he loved back in central LA, but Jamie couldn't have imagined a better night.

It had been like attending a great lecture. It reminded Jamie of his university's motto: *Rerum cognoscere dausas*, which means, "To know the causes of things." After dinner, watching Eugene race away from the restaurant in his Tesla, Jamie had turned to his companion in awe. "We just had dinner with freaking Einstein."

The artist returned to Portland, able to sleep and think more clearly than he had in years.

He continued to process all that he had learned from Eugene. He was also creating an internal Rolodex of the friends he needed to call and talk into getting the procedure. There were a few people he knew who would immediately have their lives changed if they could get out from under the traumas of their past.

However, after a couple of days, the left side of his neck—the side that had not received an injection—was noticeably pulsing. There was a crackling up his neck and down his back.

According to Eugene, the pain receptors on the right, where he'd had the shots, had achieved peace from the constant drip of tension, and now Jamie's body had become cognizant of the discomfort still being endured on the left. Eugene explained that the first doctor to try SGB on the left side was a former Navy SEAL, Dr. Sean Mulvaney. He'd found that some patients who didn't have much relief after an injection on the right benefited from SGB on the other side of the neck.

After years of clinical observation, Eugene believed that those who experienced trauma at ten years or younger found more relief from treating the left side than the right, leading him to believe that the left bundle of ganglion nerves was attached to the part of the brain affected by long-term childhood trauma. So, when the injections were applied on the left, childhood trauma was addressed. This innovation by Dr. Mulvaney allowed for a refinement on Eugene's DSR that brought important and hugely beneficial outcomes for many, many patients.

A *Neuropsychopharmacology* report published in 2015 supported this, detailing "evidence that cortical differences are seen in childhood PTSD early in the development of this illness. These brain structures are also associated with the successful attainment of age-appropriate emotional regulation and decision making."[31]

Jamie had initially believed his symptoms couldn't possibly be related to his childhood trauma. He felt he'd dealt with and overcome the past. Because of this belief, he'd chosen to have the procedure done only on the right side, which seemed to be more immediately applicable to adult trauma.

Weeks later, Jamie flew back to Chicago to be treated on the left side. It was this round of injections that really changed everything for him.

EPIPHANIES AND THE BIGGER PICTURE

The night following the artist's second procedure, Dr. Lipov called his hotel room. "It's been a few hours, Jamie. You feeling good or what?"

Jamie laughed. He felt great. Amazing. Calm but invigorated. Alive. He barely recognized himself. *I mean, I actually just laughed out loud.*

"Let me tell you why I know this is real, Eugene."

"Let's hear it."

"I grew up in the dirtiest of slums and we had these hustler dudes. You know the kind I'm talking about, sizing up everyone who walks past to see if they're a mark, someone easy to victimize. Today, my friend and I were walking back to our hotel and two street hustlers, very obviously hustlers, were shuffling around, giving me that eye. Now, normally, that kind of shit will trigger me, send me right back to being an angry kid, and then I'll be pissed off since I worked so hard to get away from that environment. But today, I'm looking at these dudes and all I can see is their sympathetic. These guys are stressed; I can see their overactive sympathetic nervous systems. Suddenly, my anger is gone and all I can feel is a massive swell of compassion."

Eugene's grin was evident through the phone. "Ex-f-ing-xactly! You get it."

"I just don't understand how more people don't know about this," Jamie said. "Why the procedure isn't more widespread for regular people."

"Eh, we try," Eugene said. "It takes a long time to change the minds of the science community. It's an old story, but look at the first guy who said doctors should be washing their hands before surgery. Around 1850, Ignaz Semmelweis, after observational studies in his hospital, saw that hand washing saved lives. Yet his germ theory wouldn't get traction for two more decades. At the time, he offended the medical community so much he was thrown into an insane asylum, where he was beaten and died two weeks later. It was much easier to believe in the status quo, that bad smells were the bad guys causing infection in surgical patients.

"Now, we take it completely for granted that it would be insane to operate on someone without washing one's hands. The overactive sympathetic is the source of a similar public health crisis around the globe," Eugene told Jamie. "But we have yet to acknowledge the danger on any real level, much less acknowledge that we can do something about it. Eventually, *everyone* will be resetting their sympathetic nervous system to the pre-trauma state whether they experience war, sexual assault, or carry the stress load for too long of emotional neglect during childhood. I mean, whaaaaaat?! I hope I'm alive to see it."

"I do, too," said Jamie.

Thinking about Eugene's caveman-and-tiger analogy, and how the fight-or-flight mechanism had been necessary for the human race to survive, it became obvious to Jamie that the symptoms were once actually *survival mechanisms*. And these mechanisms were a uniform reaction in every single human dealing with trauma, no matter what type of trauma, or otherwise the human species would have had a hard time moving forward.

What was once a mechanism biologically in place to ensure the survival of the human race is now totally unnecessary as a baseline, and it is causing widespread mayhem, despair, and death in our society, he thought.

Dr. Eugene Lipov could turn so much of that around with a simple injection.

Jamie furrowed his brow. *How much would change in our world if every human had access to this kind of relief? Eugene's dual sympathetic reset could completely change how humans move through the world.*

CHAPTER FOUR

Who, Me?

Seventy percent of humans will experience a traumatic event in their lifetime, as mentioned earlier. Eugene estimates that 30 percent of the population is currently suffering from post-traumatic stress; other sources have only slightly higher or lower numbers. This is a big deal because the symptoms evinced by many patients are harmful, emotionally and physically, not only to the individual but also to their work productivity, their social interactions, and their close relationships.

And often the trauma of others becomes our trauma and can send us into sympathetic overdrive (called *secondary post-traumatic stress*). Since most of us only associate trauma with the extremes, and since statistics can't be kept on people who are unaware of these symptoms, the percentage of the population with this overactive sympathetic nervous system could very well be astronomical.

The real issue, according to Jamie, is that roughly only 10 percent of that 30 percent (or more) are aware of the internal battle they are fighting, a battle that is spilling out into the world.

If he hadn't gone to a counselor or been diagnosed, Jamie would likely never have recognized PTSI in himself, or done anything about the long-term debilitating symptoms he had assumed were just part of who he was as a person. "I thought everyone had nightmares. I thought everyone kept an eye on doors or windows at all times. Who knew?"

Thankfully, Dr. Eugene Lipov knew.

WITH A LITTLE HELP FROM OUR FRIENDS

After he returned home from the second round of the dual sympathetic reset, Jamie was a changed person. He felt he could reach right into the massive abstract painting on his living room wall and grasp the colors in his hand. Music made his chest thrum, gave him joy on multiple levels. He slept for a solid eight hours consistently for the first time in his life.

Jamie knew who he had to talk to next. His friend Michael Thomas was a good man, a normal, hardworking, loving family guy who used his skill and reputation as a golf course agronomist to build a solid life for himself and his wife and children . . . and yet he exhibited a number of symptoms of trauma. He was by no means violent, but he was always anxious, always on alert, never able to fully enjoy a moment.

Kind and thoughtful, he was one of Jamie's favorite people. A week after his return, Jamie gave him a call. "Michael, I think this can help you."

Michael responded without a second of thought. "What are you talking about? I'm fine."

DO YOU HAVE POST-TRAUMATIC STRESS?

As mentioned, up to 90 percent of those who suffer from post-traumatic stress are like Michael and Jamie, unaware that what they are dealing with is anything but normal or typical "issues." Chronic fatigue, irrational anger, constant anxiety, or sleeplessness may seem like a way of life. Choosing to isolate is common. Thoughts of suicide crop up in the worst cases.

These responses are not normal. They are not healthy. They are not something to ignore or marginalize.

Our society has been weighed down by this pandemic of suffering without knowing it, or at the very least, without acknowledging it. Especially not by the average person, who wouldn't "dare" compare their pain to that of a warrior or a survivor of horrific abuse.

Even Marie Colvin, a prize-winning foreign war correspondent who lost an eye to shrapnel and witnessed children torn apart by bombs and personally uncovered mass graves and reported visions of a dead teenager in her bed to a psychiatrist, very famously said that she was not a soldier and so did not have post-traumatic stress disorder.[32]

When physicians see a patient who has symptoms of post-traumatic stress, they often use the PCL-5, a version of the PTSD checklist, to determine if they actually have the condition. There are versions of the checklist for both civilians and the military. A score of thirty-one or higher suggests the patient may have PTSI, though this self-assessment is not meant as a stand-alone diagnostic tool. Ideally, a physician will have a patient take the test in a clinical setting, which allows the doctor to gather more information than this simple checklist can offer. The questionnaire is included here as a jumping-off point, to help you make a decision on whether or not you should see a doctor.

PCL-5

Instructions: Below is a list of problems that people sometimes have in response to a very stressful experience. Please read each problem carefully and then circle one of the numbers to the right to indicate how much you have been bothered by that problem in the past month.

IN THE PAST MONTH, HOW MUCH WERE YOU BOTHERED BY: (0 = No 1 = Somewhat 2 = Moderately 3 = A lot 4 = Extremely)					
1. Repeated, disturbing, and unwanted memories of the stressful experience?	0	1	2	3	4
2. Repeated, disturbing dreams of the stressful experience?	0	1	2	3	4
3. Suddenly feeling or acting as if the stressful experience were actually happening again (as if you were actually back there reliving it)?	0	1	2	3	4
4. Feeling very upset when something reminded you of the stressful experience?	0	1	2	3	4
5. Having strong physical reactions when something reminded you of the stressful experience (for example, heart pounding, trouble breathing, sweating)?	0	1	2	3	4

6. Avoiding memories, thoughts, or feelings related to the stressful experience?	0	1	2	3	4
7. Avoiding external reminders of the stressful experience (for example, people, places, conversations, activities, objects, or situations)?	0	1	2	3	4
8. Trouble remembering important parts of the stressful experience?	0	1	2	3	4
9. Having strong negative beliefs about yourself, other people, or the world (for example, having thoughts such as: I am bad, there is something seriously wrong with me, no one can be trusted, the world is completely dangerous)?	0	1	2	3	4
10. Blaming yourself or someone else for the stressful experience or what happened after it?	0	1	2	3	4
11. Having strong negative feelings such as fear, horror, anger, guilt, or shame?	0	1	2	3	4
12. Loss of interest in activities that you used to enjoy?	0	1	2	3	4
13. Feeling distant or cut off from other people?	0	1	2	3	4
14. Trouble experiencing positive feelings (for example, being unable to feel happiness or have loving feelings for people close to you)?	0	1	2	3	4
15. Irritable behavior, angry outbursts, or acting aggressively?	0	1	2	3	4
16. Taking too many risks or doing things that could cause you harm?	0	1	2	3	4
17. Being "super alert" or watchful or on guard?	0	1	2	3	4

18. Feeling jumpy or easily startled?	0	1	2	3	4
19. Having difficulty concentrating?	0	1	2	3	4
20. Trouble falling or staying asleep?	0	1	2	3	4

(Source: The National Center for Post-traumatic Stress Disorders)

If you've just scored higher than a thirty-seven and still think it is ridiculous that you, an average Joe or Jane, could have PTSI, then consider this:

Trauma is not a competition. Pain is not a competition.

Trauma is trauma, slow and steady, momentary, horrific, catastrophic, or gory—what it does to the invisible machine inside each of us, controlling every system in our bodies, is exactly the same.

A thirty-year-old man falls off a ladder and breaks his leg. It hurts. He knows if he doesn't have it set and wrapped in a cast, it'll ache for a long time, and he could end up with a limp or a messed up spine. So, he goes to the doctor. If, on the same day, a firefighter falls two floors in a burning high-rise trying to save a child, and doctors are struggling to pin together her shattered spine, the man with the broken leg isn't going to think, *Oh, damn. I'm not going to the doctor. I am not nearly as injured as the heroic firefighter.*

The nurse at the ER isn't going to say, "What the hell, Bill? Stop being a baby."

Everyone's pain is individualized and just as valid as the next person's, no matter the circumstances. The doctor is going to gauge the level of care needed, do it without judgment, and move on.

Bill and the firefighter have visible wounds. Physical injuries that everyone knows need to be fixed, otherwise their physical well-being will suffer and there will likely be complications in the future. At issue here is that the negative physical change caused by trauma cannot be seen. The idea that post-traumatic stress is real has been medically accepted for decades, to a certain degree, and that fact has finally found traction in the wider public perception—though acceptance is usually reserved for combat soldiers or victims of natural disasters or long-term sexual abuse. And all of that is true.

But it is also true that trauma, like pain, can build up, cause physical damage to the sympathetic nervous system, and make life hell. Physically as well as mentally and emotionally. Allowing the brain to stay on high alert doesn't just cause invisible symptoms that affect relationships and performance; it also leads to tangible, chronic physical ailments, like inflammation of the joints or heart disease. The sympathetic nervous system directly impacts every system in our bodies: lymphatic, adrenal, hormonal, endocrinal, and neural. This will be covered in depth in chapter seven.

Public perception around PTSI needs to shift, as Frank Ochberg demands, and the medical community also needs to catch up and acknowledge the biological impacts, as Eugene Lipov is proving. The reality is that many deal with physical pain and mental anguish regularly and yet don't have to.

A REGULAR GUY

Michael Thomas's PTSI did not come from a major event or living on the front lines. His story wasn't necessarily huge, but it was human and hard. He was someone who would be considered a regular guy by his neighbors. He was financially stable with a lovely family—if not a bit annoyingly handsome, Jamie liked to joke.

Michael's trauma started with a childhood of long-term "benign" neglect, beginning the day he was born, when his biological father, the son a wealthy entrepreneur, left his wife and newborn at the hospital to be with his mistress. His mother worked hard to care for Michael and his older sister. She did the best she could with what she had, moving close to her family for help. She remarried, and when Michael was in middle school, they moved to a new state for his stepfather's job. Within ninety days of the move, the stepfather he had come to love as a father discovered he had terminal cancer and died a few months later. The next man his mother married was intense and often hard to live with. As soon as Michael acquired a driver's license and a car, he spent every summer working and living away from home. His older sister turned to hard drugs in high school. Over the years, Michael's biological father would occasionally call, making plans, inviting Michael on trips, or offering gifts that always fell through at the last minute.

Michael insulated himself by never reacting; he forced himself to remain stable, calm, and predictable no matter how he felt or what was happening around him. In the face of instability, he constructed a facade of inner strength, and he became the one who the family relied on to solve problems and make things happen. He carried a heavy burden of responsibility starting in elementary school, becoming anxious and hypervigilant, living with a persistent sense of unease. Marrying at a young age, hoping a stable, secure relationship with someone he could count on would bring something assured and certain to his life, it was a bitter blow when his first wife left him with no perceived warning, deepening the unease he carried. Then, his father died from an accident stemming from an overdose, leaving him to deal with regret and sadness, as well as guilt for feeling partially relieved.

As an adult, Michael appeared to be the embodiment of the great American Dream, having worked himself up the ranks rather quickly in the golf industry and living in a beautiful home with a beautiful family; he strove for perfection in his profession and was well respected for his stability, appearing never to be fazed by his demanding job. His extended family continued to lean on him to make the big decisions. A slow drip of low-level trauma and stress continued to build in him, year after year. When he and his new wife had children, his sense of responsibility and the desire to provide the loving environment he'd craved as a child left him in a nonstop agitated state.

"I felt like I was living at DEFCON One every day," Michael later told Jamie. "I took everything seriously, and nothing was ever good enough." Then, during his fortieth birthday weekend, while he was traveling with his wife and friends, Michael's grandmother, the stoic, unshakeable matriarch of his paternal family, passed away. Michael was faced with the thing he could not control—death—on the very weekend he was celebrating his life.

His allostatic load had become too heavy. His fight-or-flight mechanism had been on for far too long. His brain couldn't back off.

ACKNOWLEDGING SOMETHING IS WRONG

Michael is the small business owner introduced in chapter one, the one who was young and fit and still ended up at the hospital with a racing heart and chest pain not long after his grandmother passed away. Barely making it into the ER, he passed out and his heart stopped due to an extreme panic attack.

Regaining consciousness after only a few seconds, he had been surrounded by a flurry of nurses and doctors. Yet, despite the chaos, he was preternaturally calm. His body was at ease; he was not in pain. The doctors were amazed: Michael's vitals had all reset to normal levels when his heart restarted.

His body had shut down and he had awakened with a stillness and clarity of mind he couldn't remember ever having felt before.

He had just spoken to Jamie a week before and heard his story of finding peace. Driving away from the hospital later that night, he contemplated the connection. If his own body's method of "resetting" had recovered his heart rate and blood pressure and tapped out his anxiety in that brief moment, then what could Dr. Lipov's reset do for him? Michael's natural tendency toward skepticism was pushed aside. Rather than his normal, highly critical response, he found himself feeling . . . hopeful. Maybe there was a way to hold on to the peace he had found.

The trip to Chicago's Stella Center was canceled twice, but Michael and his wife, Shaw, were determined to make it happen. Finally, during the early spring storms, Michael's mother and sister flew in to stay at home with the kids for a few days. He scheduled two appointments with Eugene, so he could have the right-side injections one day and the left-side injections the next, with both Shaw and Jamie there for support.

IMAGINE

In an interview about his experience with the procedure, Michael said, "Yeah, that first day, I should have let them knock me out, but I was being the tough guy. Jamie went without the twilight anesthesia, so, you know . . . It didn't hurt, but there was this pressure and it was awkward. The stuff is injected and hits the ganglia and numbs it. I literally could feel a rushing sensation, from the tip of my toes to the top of my head. I could feel it radiating, and I was like, whoa, what is that?"

Michael's procedure was over quickly, with no side effects besides the expected numbing on the right side of his face. Besides slight sinus pressure, there didn't seem to be any other outcome.

"I was considering not going back for the injection on the left side the next day," he recalled. "It seemed like nothing was really happening."

He lay in his hotel room the rest of the afternoon, unsure what he was supposed to be feeling—it wasn't relief yet. Shaw asked him if he still wanted to join Eugene and Jamie for dinner. It wasn't until Michael got up and showered that he started to feel something new. It occurred to him as he was getting dressed that one side of his body was now peaceful, while the other was, by contrast, a live wire, buzzing and erratic. He realized that the side at rest was the side on which he had received DSR that morning. Intrigued, he told his wife he was still on for dinner.

A few hours later, Eugene pulled up to their hotel in the Tesla, chauffeuring Jamie, Michael, and Shaw to the restaurant. "Michael, how are you feeling?" he asked.

"My left side feels like it is going a million miles an hour, while the right feels calm and steady, like it's going two miles an hour."

Back at the Stella Clinic the next morning to perform the injections on the other side, Michael opted for the twilight anesthesia, meaning he was out when the injections were made on the left side of his neck. The doctor was there when he woke up in post-op.

"How do you feel?"

"Ummm . . ." He focused. "Amazing! I feel so good! I don't know how to describe the peace to you, Dr. Lipov, but it's amazing. That's just the best way to describe it."

While Michael dressed, Eugene went to speak to Michael's wife, who was with Jamie in the hallway. When the doctor told her that Michael had woken up and said he felt amazing, Shaw surprised him by bursting into tears.

"You don't understand. My husband has never said that. He never feels *amazing*," she said.

A smiling Michael joined them just as a young woman walked into the lobby of the surgical center, bounded over to Lipov, and enthusiastically embraced the doctor. After a long second, she turned to the group, her eyes shining above her COVID mask, and she said, "This is a miracle man."

Michael and Jamie nodded, fully in agreement. Jamie then said to the doctor, joking, "She must work for you; how much are you paying her?"

After she left, Eugene teared up as he told the group that this same girl had come in for the procedure the previous day. "Yesterday, she couldn't have me in the same room with her, much less touch her! She's a survivor of sexual trauma—probably the worst sexual assault victim I've ever treated." Now, the day after receiving DSR, she was a different person, casting off a glow of energy that had replaced the fear and hyperanxiety.

The air outside was chilly and damp, a drizzle that threatened to become more. To Michael, what could have been just another gray winter day in Chicago offered bright silver skies and shimmering drops hanging from the tree branches. He found beauty everywhere he looked, even in the wet, shiny concrete. Twenty feet from the doors of the Stella Center, Michael said to Jamie, "If this is the only reason I was ever meant to know you, it is because of what just happened to me."

While Michael was resting at the hotel, Shaw and Jamie stepped out to grab some air. They stopped at a raw food café and perused the grab-and-go items. The woman behind the counter, the owner, surprised them when Jamie casually asked how she was doing and she responded, "Okay. I mean, it's been a year. Our front window was smashed in twice during the riots . . . and my ex-husband killed himself by jumping off a high-rise." What came next stunned them even further. "The good thing is, my daughter was there when he did it."

The two glanced at each other with disbelief. How could it be a "good thing" for a child to see her parent end his own life? And did a stranger really just share what felt like an ordained conversation?

The owner went on to explain how her daughter had seen in her father's eye that it wasn't actually him in there anymore. His eyes were empty; what made him *him* was gone long before he leapt to his death. That awareness gave the daughter closure.

Leaving, Jamie and Shaw couldn't help but compare their own cases with the man who had just jumped. Their trauma had not risen to the level of suicidal ideation, for which they were grateful. And they'd had access to Dr. Lipov and his procedure. If this man had known about DSR, he would likely still be here.

Later, Michael put a hand on Jamie's arm, very serious. "Imagine if everyone did this. I feel like I have a second chance at life. Imagine a city full of people who could have this kind of relief. This peace."

Jamie closed his eyes. "Yes. What would that world look like?"

This experience stuck with Jamie. Beyond his own relief, and the relief of his friend, there was the idea that huge populations could benefit from, first, coming to understand the post-traumatic stress injuries within themselves, and, second, knowing there was a medical treatment that could alleviate their pain and suffering. That could stop anxiety and feelings of doom. Stop violent impulses or suicidal ideation. Bring lightness into their dark.

What can I do? Jamie thought. *How can I get the word out?*

CHAPTER FIVE

The Unlikely Avengers

IT'S ALREADY WORKING

At every intersection, Jamie was becoming increasingly focused on PTSI and how it was affecting people and society. Soon, his other projects were suffering as he did deep-dive research and had long conversations with Eugene and others.

It was astounding how many people fell into Jamie's path around this time, people who soon were just as fascinated by the possible far-reaching implications of DSR.

For instance, when Jamie's friend Dr. Jay Faber heard about the outcome of Jamie's procedure, he insisted that Jamie have a conversation with Faber's mentor, Dr. Daniel Amen.

Dr. Amen agreed to a Zoom call. Peering intently at another computer screen, the doctor talked to Jamie about the peer-reviewed journal articles on DSR and Lipov. "Huh. Interesting. I'm looking at a credible study. I'm looking at these numbers from Womack, and they're saying the efficacy of Lipov's procedure is around seventy percent."

"Yeah, I know, but those are old numbers," Jamie said. "They're still just doing the one shot on the lower right side at Womack. Dr. Lipov has been doing two injections on both sides, using an ultrasound for accuracy. This really is working."

"Jamie. Even at seventy percent, these results are very promising."

Jamie paused. "Oh. Okay. But seriously, at the Stella Center, they are up to eighty-five percent efficacy now, and the results are lasting years, if not permanently."

He and Jamie talked about the scans he used to determine the details of his patients' traumatic brain injuries and how that might apply to Dr. Lipov's procedure. Furthermore, Jamie thought maybe the story of his journey could be helpful to Eugene, who struggled with growing a platform in the public sector.

"Dr. Amen, you said you took crap in the early years for your use of brain scans, when you started in 1989. Why is that?"

"In every area of health, we take an X-ray if we think there's a problem. Yet, when someone is deemed 'crazy,' we historically don't look at the brain. Why not? In the past, people tried to label me as a quack, but, over time, neuroscience has matured, with more research and data indicating that my theories are accurate. I didn't give up."

It was true. He now had a number of clinics across the US and his SPECT brain imaging had transformed how psychiatrists were practicing today. The single photon emission computed tomography was a nuclear imaging technique that used a radioisotope, technetium, injected into the bloodstream to measure blood flow. Amen used the SPECT to understand strokes, epilepsy, trauma, some types of dementia, drug use, and more. With that knowledge, he went on to provide more targeted care.

When Jamie asked how he got started, the doctor told Jamie a story about his nephew, aged eight or nine, getting kicked out of school for violence, and his sister calling Amen in a panic. They scanned his brain, only to find that the boy had a golf ball–sized cyst in his head. They drained the cyst and his behavior went back to normal.

"He's in college now, completely fine. If I hadn't scanned his brain, he would have ended up in the criminal justice system."

Amen continued to scan brains, well aware of the importance of his work. He now has a massive data set of more than 190,000 scans. "Jamie, thirty years later, people are still coming to me, telling me if I can't help them that they're going to take their life. I can't turn my back on that. I still feel the weight."

Amen's compassion for his patients was touching. Jamie cleared his throat. "Dr. Amen, this DSR . . . it works. I did it. It's completely changed my life. It's changed my friend's life. I've been talking to other people; it's working for them. You should talk to Dr. Lipov."

Both Dr. Amen and Dr. Lipov work on the principle that injuries to the brain can be reversed. Amen was intrigued.

Jamie left the Zoom meeting feeling energized, but not really sure what to do next. What else could he do to help move the needle on PTSI public awareness? Luckily, the universe continued to intervene.

THE ASSHOLE SHOT

During this time, despite his new interest, Jamie's book *The Iconist* was getting noticed. As part of a growing speaking career, he was invited to a number of events to talk about branding and leadership, including a podcast with hosts Professor Jared Nichols and Lieutenant Colonel Paul Toolan. He was asked to join a discussion on taking an active role in shaping the post-pandemic future.[33]

Jamie made a connection with Toolan, who spent his days as a commander for the military's Special Operation Forces, working out of Fayetteville's Fort Bragg, one of the largest military bases in the world. A week or so after the podcast, Toolan reached out and asked Jamie to come to Fort Bragg and speak to Special Forces and the Psychological Operations division (PSYOPS) about his book and being an influencer. The last author to speak there had been Simon Sinek, the author of *Start with Why*, *The Infinite Game*, and *Leaders Eat Last*, company Jamie found flattering. Toolan admitted the pay wasn't great, but Jamie would be speaking at the JFK Auditorium and would have the DOD as a client.

Paul Toolan was a man heavily invested in combating trauma in soldiers, as well as other impacted populations. His background was impressive; he'd led Special Forces teams in Afghanistan and the first Special Forces Battalion in the Counter-ISIS campaign in Syria. He was currently the deputy commander for First Special Warfare Training Group, responsible for safeguarding and leading an organizational culture designed to select and train Green Berets. And he established military-based organizations geared toward helping soldiers and their families.

Once retired, Toolan planned to work full time to help soldiers safely transition out of the military and back into civilian life and family. He started JANUS, a program solely dedicated to this kind of transition care and destigmatizing what he calls "the demoralizing mental health label on our nation's warriors." The program's name comes from the Roman god Janus, always portrayed with two faces and associated with gates, beginnings, doors, and passages. As Toolan realized that post-traumatic stress was a fixable "injury" and not a disorder, he saw a huge opportunity to restore dignity to returning soldiers and their families.

Jamie's initial meeting with Toolan on the podcast—completely unrelated to PTSI—was fortuitous. Before calling Toolan back to further discuss the logistics around the trip to Fort Bragg, Jamie mentioned the Special Forces connection and the invitation to Eugene, who was now a friend he spoke with frequently.

"I know they're doing SGB at Womack," said Eugene. (As noted earlier, Womack Army Medical Center is located on Fort Bragg.) "Maybe you can ask your Special Forces contact what kind of results they're seeing. I haven't been able to find out much about how they're using it."

When Jamie explained his growing interest in post-traumatic stress and Lipov's work to Toolan, the lieutenant colonel did look into it. He found that Womack was performing the procedure more than a thousand times a year. Toolan brought Jamie onto a Zoom call with Master Sergeant Geoff Dardia, the head of Third Special Forces Group Human Performance and Wellness (HPW) and the founder of Task Force Dagger's Special Operations Forces Health Initiative.

Within a few minutes, Jamie realized the Womack doctors were using an older version of Lipov's procedure (closer to the original SGB), performing only one injection on one side. Yet, even with the old methodology, they were getting 70 percent efficacy, though the positive effects were often wearing off within a year or two. If the procedure isn't done with the modern and full protocols, regression is much more likely, and the relief is substantially less.

The spouses of the Special Forces members who were undergoing SGB at Womack called it the *Asshole Shot*. They said they knew it had started to wear off when their husbands went back to being assholes. They'd urge them to get the procedure again because of its astounding effect to bring them back to who they were before the ravages of war.

The SGB/DSR had recently been reported as a successful treatment of trauma and PTSI symptoms by four military institutions (Walter Reed National Military Medical Center, Long Beach, California; Tripler Army Medical Center, Honolulu, Hawaii; Naval Medical Center, San Diego, California; and Womack). Soon, Jamie and Eugene hoped, doctors would be using the DSR improvement across the board.

GRAND ROUNDS

Jamie got permission to bring Eugene onto Fort Bragg as a guest during his *Iconist* lecture. Toolan was also able to schedule the pain physician into grand rounds at Womack Army Medical Center during that time, an appointment that usually takes more than a year to set up, if it happens at all. Lipov loved the arrangement, which would give him the opportunity to go into the clinic and talk to the doctors about adapting to the DSR versus SGB.

Six weeks before the book lecture, Jamie traveled to Fort Bragg to hammer out details and ended up speaking with Special Forces Psychological Operations. Toolan also scheduled extracurricular meetings that included riding on a Black Hawk helicopter while hanging from a rope with a bunch of trainees, dinners with Special Forces and their spouses, and—after he mentioned he was kicking around the idea of a documentary on PTSD versus PTSI—a face-to-face meeting with Dardia and a group of combat soldiers dealing with post-traumatic stress injury. Jamie suddenly felt like Kelly McGillis's character in *Top Gun*, an outsider with a front-row seat on a military installation.

As Jamie sat down to dinner with Toolan, he was introduced to Senior Technical Advisor Steve DeLellis and his wife. Steve mentioned he'd heard that Jamie had "had a fun morning."

Jamie excitedly told him how he'd dangled from the big military lift helicopter. "It was like I was in *Black Hawk Down*," the artist said, laughing.

"Yeah?" Steve replied. "I went down in the first chopper."

It turns out, Steve was an extremely well-known Special Forces guy and had been in the seventeen-hour firefight in Mogadishu. Before Jamie could swallow his embarrassment and apologize, the retired Special Forces warrior jumped into storytelling mode. He described how on that terrible day in Somalia, another soldier who was in the crash had turned to ask Steve, who was a medic at the time, what they should do. Steve had looked at the guy—whose nose was hanging from his face—and replied, "Start shooting!"

Listening to the stories from the men around the table, Jamie thought of the reality they moved through, the depths of suffering with which these guys dealt. At every turn, he was being brought into contact with more and more individuals well respected in the military community, a community that had been dealing with post-traumatic stress more openly and aggressively than any other sector in the United States and so had an invaluable base of resources and knowledge.

FORT BRAGG

The next morning, Jamie arrived at Fort Bragg; and he found his time there to be a master class in why soldiers are admired.

Driving onto the base, he was amazed at the size of the massive complex. The army's Fort Bragg covers roughly 250 square miles in North Carolina—the largest military installation in the world—and hosts the world-class Womack Army Medical Center, as well as 54,000 military personnel, including Airborne, Special Operations Forces.[34] The cleanest, most polite "city" a civilian could ever hope to visit.

Master Sergeant Geoff Dardia arrived promptly for the Monday morning meeting set up by Toolan. A career Special Forces soldier, Dardia was pleasant but distant—the lifelong military man had no reason to trust Jamie, a civilian author.

Though humble, Dardia was a guy everyone respected. He ran the Health Initiative Program with skill and was considered a no-nonsense yet forward-thinking military healthcare provider with a focus on a whole-person approach. He was also the brain injury and PTSI expert at Fort Bragg; many among the Special Forces believed he knew more about military health and post-traumatic stress injuries than most doctors.

Dardia brought Jamie into a meeting room, where ten to fifteen Special Ops folks were waiting to discuss their post-traumatic stress experiences after multiple tours of combat duty.

A two-and-a-half-hour session ensued. "A round-robin of trauma and crazy-ass stories," according to Jamie. For example, a young Special Forces operator was carrying around this memory: His squad, using military dogs, had chased a guy into a building. As the combatant was about to enter an apartment and possibly escape, they let the dogs go. The door opened and the terrorist reached in, grabbed his seven-year-old daughter, and threw her to the attacking dogs so he could escape. She was torn apart.

The soldier knew something in his mind was triggered, that he couldn't find a way to respond to the world around him like a normal human anymore. He and the others in the room were pissed off and trying to find help, but people would refer to them as crazy, and therapists would try to counsel them on finding willpower. It filled Jamie with rage, hearing about the physical, mental, and emotional sacrifices these people had chosen to make for this country only to be ignored or belittled in their time of need, even by some in the military community. And their need appeared to be great. They came back with gruesome memories, survivor's guilt, broken bodies, brain injuries, and limited support outside the military community. At the very least, Jamie thought, they should not be stigmatized. It should be acknowledged that they had a biological "injury" and not a disorder. They deserved that much—as does every other human, no matter how they move through the world.

A human can only carry so much stress, and Special Forces operators were asked to be superhuman in this regard. These soldiers had an ever-growing allostatic load, and that buildup of chronic stress eventually began to affect physical and mental well-being.[35] In military circles, the more time in combat, the higher and more health debilitating the load.

Dardia used the term *operator syndrome* multiple times during the meeting. The term referenced how when these warriors left the battleground, they found they couldn't stop "operating" like a combat soldier. According to B. Christopher Frueh in the *International Journal of Psychiatry in Medicine*, a well-respected research and medical journal, "Operator syndrome may be understood as the natural consequences of an extraordinarily high allostatic load; the accumulation of physiological, neural, and neuroendocrine responses resulting from the prolonged chronic stress; and physical demands of a career with the military Special Forces."[36]

Every person in that room was struggling with a litany of symptoms matching Frueh's list. And from that list, Jamie at first concluded that operator syndrome was a manifestation of PTSI specific to military Special Forces.

Then it hit him like a two-by-four upside the head. He knew people from his past who had obvious operator syndrome but were not soldiers.

CONNECTIONS

The top six most prevalent symptoms of operator syndrome, which he heard again and again from the Special Forces operators that day, were the same PTSI issues that many people had dealt with in Jamie's childhood poverty-stricken neighborhood—anxiety, rumination, sense of doom, hair-trigger anger, hypervigilance, and lack of sleep—and continued to deal with even after removed from the distressing environment.

Further, an overactivated sympathetic nervous system also causes adrenal, lymphatic, hormonal, neural, and endocrine problems. Which leads to the following physical issues: headaches, orthopedic problems, cognitive impairments, vestibular impairments, sexual dysfunction, hypertension, heart disease, even cancer. And any of these "mental" symptoms can be exacerbated by concussive brain trauma experienced through war, or by any prolonged allostatic load that lasts too long.

These same stress loads and resulting mental and physical ailments are very common in inner-city neighborhoods around the world where people are struggling just for food, safety, and a place to sleep. But even after escaping the inner city, many continue to "operate" as if they are still in duress.

As he sat with the soldiers, another fascinating, more surprising connection occurred to Jamie. His head spun. He flashed on past research he'd done on the incarcerated while writing *The Iconist*. He'd been curious about prolonged isolation and its effects on the body. He realized now that what was being defined as operator syndrome in military veterans and Special Forces war heroes was also a primary symptom among the incarcerated. What if the sympathetic overdrive was causing overreactions and driving crime? What if this condition was made worse inside prison? And it was not only why recidivism was so high, but was it also creating a public health threat by making communities even more dangerous when people get out? If laws are based around intent, why are courts imprisoning people who are being duped by their stellate ganglions?

As the group took a quick break, Jamie scribbled notes furiously. He'd always been interested in prison populations and what happened to people's minds behind bars, and so he'd done light digs into behavioral research in the past. Jamie thought of Faber's book, which had covered brain health issues in prisoners. Faber was the nation's leading authority on brain health and crime; he traveled the country serving as a professional witness, providing documented research to judges and prosecutors that details how poor brain health can at times nullify any concept of intent. Some prosecutors and judges are starting to listen.

His mind was suddenly on fire, amplifying all kinds of correlations.

His thinking went like this: In the military community, where they were trained to protect the innocent while surrounded by enemies, the feeling that the "tiger" was going to kill them at every moment of every day eventually manifested itself as suicidal ideation. In other words, if you knew you were about to die at any second, you became resolved to die.

In the inner city, where there was also a deadly tiger perceived around every corner, ready to pounce at any second, the same mechanism was at play, telling the amygdala that even the slightest argument or show of disrespect could be deadly. In this environment, however, the reaction to perceived danger could manifest itself as homicidal ideation.

When two armed gangbangers in the inner city got into a minor argument, their sympathetic nervous systems "lied" to their amygdalas, insisting that the confrontation was a threat with a life-or-death consequence. Confrontations generally ended one way: one of the gangbangers was going to the hospital or the morgue, and the other was going to prison. Because of an allostatic load that had been building over a lifetime of poverty and crime, and the ensuing biological impact on their sympathetic nervous systems, their morality or desired intentions had little or nothing to do with their actions in that moment. Logical behavioral responses simply weren't available.

Could this broken invisible machine, the overactivated sympathetic nervous symptom, be a primary cause for impulsive violent crime?

Jamie filed that thought away for later.

Operator syndrome, a consequence of the overactivation of the sympathetic nervous system, explained how a felon would continue to operate as if he was still living in the high-stress world of prison long after he'd been released. Operator syndrome also explained how someone like Jamie, who grew up searching for love and avoiding the abuse of adults every day, could move away from inner-city poverty and violence, away from the stress and the danger, but still operate like they were living back in a one-room slum apartment with five people . . .

When Jamie finished the meeting, Dardia waved him over; his demeanor had changed and a trust had begun to build.

"Hey, look at this." Dardia held out his cell phone to Jamie. On the screen were listed the six key biological symptoms of operator syndrome: anxiety, hypervigilance, sense of doom, lack of sleep, rumination, and hair-trigger anger.

Jamie's mouth fell open. He excused himself and immediately called Eugene, who was somewhere on the base at a medical meeting. The doctor picked up and Jamie said in a rush, "Hey, have you ever heard of operator syndrome?"

Yes, the doctor had heard of it. Jamie continued, "Did you notice the core symptoms are that of an overactive sympathetic? Basically what your innovation addresses?"

Eugene answered that he had never thought about it, but it made sense. Jamie's head was spinning. He was thinking of all the people he had grown up with in poverty who were likely walking around with operator syndrome, and how it was physical. The revelation changed everything. Later, Jamie would come to believe that upward of 30 percent of the civilian population worldwide could have operator syndrome. Another revelation.

Jamie couldn't believe that heroes and murderers were suffering from the exact same physical injury and symptoms—and that didn't seem to be on anyone's radar. He thought, *People need to see DSR in a different light. A light that needs to shine on all of us.*

SPEAKING TRUTH

Dardia remained polite but aloof with Jamie throughout the rest of the meeting. However, he was attentive, firm, and compassionate with the warriors in the room willing to share their most horrific experiences.

As the meeting was wrapping up, Dardia turned to him. "All right, Jamie, anything you want to add here?"

"Can I say whatever? I mean, can I be honest?"

"Why wouldn't you be?"

"Listen . . . none of you are crazy! It's crazy that someone would even let you believe that. None of you have a *disorder*. What, is this 1972? How are doctors still calling this a disorder? All these symptoms you're talking about, these are because of a damn physical injury to your nervous system."

Jamie swallowed hard, a lump of anger stuck in his throat. He was shocked that even among those who studied military-based PTSI, stress as a disorder was still a notion. *The science is out there*, he thought, incredulous.

Many of the soldiers gathered that day had already had the stellate ganglion block; however, they'd received only the one shot using the old methodology. This "Asshole Shot" did alleviate some of their symptoms for a short time, making them calmer, so it wasn't a total waste. Jamie looked around the room and hoped they'd be willing to try again, with the updated dual sympathetic reset, a quantum advancement.

"I do know some of what you're talking about, though I realize it is on a different level. I'm no hero." Jamie took a breath and told them his own story of "childhood bullshit" and how he'd come to terms with a PTSI diagnosis "as a consequence of enduring that traumatic bullshit" and finally found relief from the symptoms. "It hurts my heart that you're sacrificing so much, as well as your family, your kids, to come back and have people tell you that you're just crazy or have a 'disorder.' It's wrong."

Walking out of the meeting with Dardia, the master sergeant said, "Listen, there's someone you should meet. His name is Trevor. He's a fifteen-year Green Beret who joined the military because he thought it would be more honorable to die for his country than to die by suicide."

Dardia's voice was calm but serious, and he was finally looking at Jamie. "It's not uncommon for Special Forces applicants to come from abusive homes. Most are here because they want to make sure those who can't defend themselves are protected. Talk to Trevor. You'll see what I mean."

DI OPPRESSO LIBER

"*Di oppresso liber,*" Jamie said to Trevor Beaman a few months later. "The Green Berets' motto *to free the oppressed* . . . What does that mean to you?"

The E8 master sergeant, Special Forces, sat across from Jamie, a camera trained on him. They were in Paul Toolan's office at the Special Warfare Training Group Headquarters.

Trevor had agreed to being filmed for a documentary about Dr. Lipov's pioneering breakthrough, discussing his military career, his operator syndrome, and his childhood abuse.

The lanky soldier leaned back in the wooden chair, stretched out his legs, and tilted his head in thought. Trevor was over six feet tall, with the face of a Calvin Klein model, the brains of a guy who went to both Purdue and the National War College, and the heart of a dedicated warrior.

"I was a history major in school," he said. "It was then that I [realized] I have this ability to talk to people and gain rapport with people. [For instance], I don't look at Afghans as the enemy, they're just human beings . . . That's what matters. This idea of freeing the oppressed is giving people a voice, giving people the power to defend themselves and their family . . . If you truly are a Special Forces operator, you have to embody that within your soul. And if not, you shouldn't be a Green Beret. Your life now is to help other people with your skill set, your ability to train and lead, the ability to sacrifice your life for someone else's freedom."

Jamie interjected, reminding Trevor that the perception of the Special Forces was not always so positive. Hollywood often portrays Green Berets as testosterone-filled guys who are occupiers, not compassionate helpers. "Most people don't understand you're often there building schools, not just blowing things up."

Trevor nodded. "To say that *every* Green Beret is doing the right thing or accomplishing the right goals or being a true operator, you know, while working in austere environments, that is not truthful. It is about who you are as a person, as well. That really dictates what you do. Because when you're [in] the hinterlands of

Afghanistan—there's no rule, there's no law, there's no security, it's just what you provide that keeps you alive . . .

"When I was down in Kandahar . . . I told myself at that point, this is what I am going to do for a living. Being out in the middle of nowhere . . . being a person who *can* do better, helping the people start to defend themselves against the Taliban and other [dangerous] people outside of the village. And then there was this political thing I witnessed that changed me forever: I got to see the first woman vote in the election for President Karzai. To think that we as an army, as a nation, helped allow that to happen was powerful. It meant something. That's freedom . . . That's freeing the oppressed."

Jamie crossed his arms and forced himself to ask the next question without emotion. "Was there a connection between your sexual abuse and your desire to join Special Forces?"

The Green Beret had been open with Jamie before the official interview, telling him every awful detail about how his stepfather had begun sexually abusing him from around the time he was six or seven to the time he was sixteen. The abuse was constant and did not end until Beaman's mother found out what was happening—she worked the night shift as a nurse at the hospital, unaware of the horror her son was enduring almost nightly.

"My penis would be so raw, and I would have to feel that, and have that be a part of me, throughout the day . . . When I was in seventh grade, I tried to commit suicide for the first time . . ." To ensure the boy didn't say anything about why he'd attempted to kill himself, his stepfather remained at Trevor's side in the hospital room for the duration of his stay.

Trevor was willing to talk about the dark side of his past because he recognized that revealing the post-traumatic stress originating from his extreme childhood trauma might be just as helpful to some as stories of his combat trauma. Trevor grew up with a mom who loved him but was gone much of the time for work. Otherwise, his life was a showcase for someone apt to acquire PTSI.

It wasn't just the monstrous sexual abuse; his family lived in the hood, in poverty. Their home for years was a condemned apartment building. He had to learn quickly how to keep himself safe out on the streets, and safety was found in numbers. "I was fourth grade-ish and I got 'beat' into the Latin Kings. I felt like these were the people that gave me some type of purpose that I didn't have coming from my household . . . I wasn't at that age to be a part of [the violent crimes] but it was, like, can you go over here and vandalize this stuff? Can you just be at this park, when we have this meeting, so that if something goes down you can be a part of it? I did not look at it as gangbanging. I looked at it as another family that provided mentorship. And it was protection."

But then his mother saw the bruises from fights, and she called the sheriff. Trevor did not say a word about the gang, but he was labeled a rat. So, suddenly, his time outside of the home was almost as consistently dangerous as it was inside, home alone with his stepfather each night.

Trevor survived all that, went to college, mastered the grueling fifty-two-week Special Forces Qualification Course (only 60 percent of the handpicked recruits successfully complete the infamous "Q Course"), and became an upstanding protector of our country. But "survive" is a measured term.

As Dardia said, there are many soldiers who suffered abuse as children who then take that pain and go on to find a way to save others from the same kind of fate. A recent *JAMA Psychiatry* article supports this, saying, "Adverse Childhood Experiences (ACEs) are associated with several adulthood health problems, such as self-directed violence. For some individuals, enlistment in the military may be an instrumental act to escape adverse household environments . . . Previous research about adverse childhood experiences (ACEs) clarifies the serious public health ramifications of early-life stressors."[37]

Trevor was such a man. He was empathetic to those in need and willing to do the hard thing, whatever that entailed, in spite of or maybe because of a severely traumatic childhood.

"Do I think there is a connection between my past and joining the military?" The soldier's shoulder jerked in a shrug at Jamie's question. "The easiest and simplest answer is that I don't think I did it intentionally. I think my body just knew that's what I had to do because of . . . what I've been through . . . I knew what it felt like not to have a voice, and I didn't want that for another person.

"The Green Berets gave me an avenue [to do] that, morally and lawfully. [In the States], it's not so much that way. An army, you can do nasty things to bad people and it's all behind the flag. I'd be standing up for the voice of the people, the same people I'd spent years with, having done back-to-back trips to the same station. You get to learn and love those people, working beside them every day. As you get in gunfights, as you see people you know step on IEDs . . . it's a bond that is created in a really austere, hard environment.

"There's a lot of good that comes out of it. But . . . but there is a price that you pay for it. And that's not anything anyone can see."

OPERATOR SYNDROME

Beaman's trauma symptoms were severe. And his family was suffering the brunt of his seemingly uncontrollable aggravated responses to the world around him.

He had tried every kind of drug from around the world to dull the edge of his post-traumatic symptoms; he was also trying to erase the weight he felt at having taken human life. He wanted so badly to be as good a father and husband as he was a warrior, but his operator syndrome was making him unpredictable and dangerous at home—exactly the opposite of the kind of man he wanted to be.

His childhood years had been so intense, with years of horrific sexual abuse, poverty, and gang violence, that his allostatic load had been well past the tipping point by the time he joined the military. Hoping to alleviate his trauma by doing something good, he'd worked his way up to Green Beret, where he faced 24-7 highly intense physical demands, and death was a constant personal threat. Further, witnessing others subjected to death and violence, and oftentimes being the cause of that death and violence, led to a moral guilt that is deeply embedded in the psyches of a large percentage of combat soldiers. The toxic stress makes it hard to sleep or eat well, which becomes a catch-22, since eating and sleeping are the body's way of providing recovery resources. As mentioned, Special Forces operators, like those growing up in extreme poverty, carry a superhuman allostatic load. And Trevor Beaman more than most.

Who he had to be in order to survive his early years, and then who he had to be to survive overseas while being an effective soldier, created a man who instinctively operated with survival as the goal, consciously or not.

For years, the tiger was always around the corner, if not right at Trevor's throat, fangs bared.

Boarding a military transport, dropping down into the jump seat with a grunt, watching a smoking desert ruin recede . . . Trevor was not going to step off the plane in North Carolina, shake off the tiger threat, and morph into an easygoing, lighthearted person, totally fine with the uncontrolled civilian environment around him. Going into a crowded grocery store was not going to be a minor irritant or time suck; it was going to be a place filled with people coming abruptly around corners or waving their arms or moving too quickly, and that was nothing but *danger, danger, danger* for everyone involved.

Operator syndrome cannot be controlled by willpower or strength of character, otherwise Trevor would have kicked its ass long ago.

Trauma specialist Jennifer Satterly addresses this in *Arsenal of Hope*: "Understand that PTSD [and the resulting operator syndrome] is a *biological* condition . . . A person with a brain tumor is not blamed for the injury to their brain, or the resulting symptoms. When a cancer patient's personality changes, we blame the disease and don't expect that the patient can just 'snap out of it' without medical intervention. Rightly so. This is also accurate for our soldiers with PTSD. They did not ask for their brain functionality to be altered, nor do they have the ability to just 'snap out of it.' *Willpower, no matter how strong, cannot resolve the biological issues.*"[38]

Trevor knew that if he wanted his children and wife to be safe with him, he had to find a way to turn off that part of his brain. He had never desired anything more in his life. He knew he needed the DSR procedure.

WHAT ABOUT THE OTHER GUYS FROM THE HOOD?

After Trevor underwent Eugene's procedure, Jamie again interviewed the Green Beret. Trevor looked directly at Jamie and choked. "You know, I was a skeptic about all this. I thank you with my whole heart."

There was a pause. Everyone in the room was fighting back tears.

"After the treatment with Dr. Lipov, the world evolved into a different perspective," he said. "Before DSR, my brain was consumed by the horror film of my childhood. With the DSR taking that away from me, it did not change my identity, it did not take away my history, or my ability to be a Green Beret . . . what it has done is to allow me to let go of all the trauma and super hard memories. It's now a long-ago memory, something I went through, but no longer consuming me," said Trevor.

"I didn't have to start my day off convincing myself to live."

He went on. "Then as I started to think about my abuser while I was growing up, I had this feeling I could go back to where I grew up and shake this person's hand and thank him for the gifts that he gave me. It was the first time in my life that I didn't want to murder this person. Since then, the impact of those events and moments in my life have become a distant memory, not engulfing the day to day of who I am. That allows me to be present at my job, as a father, as a husband, and not to see the horror film all day. Makes me feel like I can be a real person. And it's incredible."

Trevor spoke about other treatments he'd tried, comparing them to DSR. He admitted when Jamie and Eugene offered him the treatment, he was skeptical.

"When I did EMDR [eye movement desensitization and reprocessing], I focused on childhood trauma, I didn't deal with a lot of war stuff. And so the feeling of what the DSR did was similar to what EMDR was like, but without all the pain. And EMDR only got me to maybe sixty percent good.

"Now, you can't just go to Eugene Lipov and be fixed. There is a lot of fucking work that has to go with it. But . . . nothing has helped me like this! The other modalities I've done were based on readjusting my behavior and how I view the world. I'd be trying to convince myself I'm okay and that I can manage my life. But I'm sick and tired of just managing my life. *The treatment with Dr. Lipov was the cure.* Not just managing, not just keeping yourself busy without fixing anything. DSR is the true cure to shut it all off, finally, and now I can start anew.

"There's been this reemergence of my innocence. I'm back to me again, the me I wasn't allowed to be. Now I can truly live. Someone had stolen that innocence from me in my childhood. I'd given up getting that back. But this has reignited it in me again. And I have a . . . a rebirth. Like the Radiohead song says, I'm truly lucky to be alive."

When asked why he, a professional soldier still active and with a reputation on the line, was willing to share his personal pain so openly with the world, Trevor's answer was forceful and heartfelt.

"Because, man, *I* will always get the treatment!" he said. "The military has spent millions on my training and will get those of us in Special Ops the best available care—but what about all the guys on the block, the guys from the neighborhoods where I grew up? What about *them*, Jamie?"

It was true. He was getting the best that was available to alleviate his symptoms because he was Special Forces. He was right to ask: What about everybody else? What happened to the kids he'd grown up with or who grew up in similar situations, impoverished and surrounded by violence with gang members and drug addicts as role models?

Trevor Beaman and Jamie Mustard were both children of poverty and inner-city violence. But they were able to channel the symptoms of their post-traumatic stress in ways beneficial to society, even if their personal lives suffered. However, a large population of those from the same type of background aren't lucky enough to experience the same interventions Trevor and Jamie had. What were those people doing with their allostatic load and the resulting PTSI?

According to Jamie, most of them will grow up with symptoms of operator syndrome, not understanding that their psyche continues to believe that they must operate at a fight-or-flight level to survive. Many will deal with anxiety, rumination or sense of doom, hair-trigger anger, hypervigilance, suicidal ideation, and lack of sleep, even if they've grown up to live in nice condos in an art district.

But now Trevor, like Jamie, had experienced relief thanks to Dr. Eugene Lipov. And he was ready to help spread the word.

COMING TOGETHER

When Paul Toolan and Geoff Dardia started talking with Jamie and Eugene about DSR, and then Trevor joined the mix, Eugene was thrilled.

But he was amazed when trauma physicians Jay Faber and Daniel Amen stepped in, offering to donate before and after scans of Green Berets undergoing the procedure.

"Jamie, how are you doing this?" Eugene asked.

Jamie didn't know how to answer. He was equally astounded at how the universe seemed to be putting the right people together, or even that these warrior guys and famous scientists were willing to bring Jamie into their circles. Sure, similar to them, he'd carried a heavy allostatic load, and they could read that in each other. That probably drew them together on some basic level. But he also believed it was the way of the artist to break through walls—walls that most don't even realize are there—and replace them with bridges. Artists connect humans at deep, unseen levels, and that was what was happening here.

Little did either of them know that soon there would also be a sheriff, an inmate, a famous war correspondent, and many others stepping in to join their efforts.

CHAPTER SIX

The Criminal Mind

FROM WARRIORS TO PRISONERS

Evidence from large military studies irrevocably links PTSI with physical health, suicide, housing and homelessness, employment and economic well-being, social well-being, aggression, and violence. And criminality.[39]

The sympathetic nervous system has no way to discern what is causing allostatic load or the type of threat it perceives. As Paul Toolan likes to say, it is "apathetic." It is mechanical. It interprets all extreme violence or stress, physical or emotional, exactly the same. Leaving soldiers and children alike to suffer post-traumatic stress symptoms, even operator syndrome.

When describing someone with operator syndrome, "you're describing someone who is highly adapted to an unsafe and chronically dangerous environment, where their basic relational needs were not met," said Robbyn Peters Bennett, LPC, a psychotherapist who agrees that the symptoms of OS are not just for soldiers. She has worked for years with children and adults who suffer from anxiety and dis-regulation and distress usually due to childhood abuse, lecturing internationally (including a TED Talk[40]) on the effects of trauma and neglect and ending family violence. She agreed to an interview with Jamie in December 2021.

"We understand soldiers," Peters Bennett said. "We get it, they're out there performing superhuman tasks that are terrifying. And even if they're not engaged in military conflict, they're waiting. Soldiers spend an enormous amount of time waiting around while remaining vigilant, with nothing happening. And it goes on for a long time. So it simulates in some ways childhood trauma, in particular a child who is raised in poverty in the ghetto or in a gang environment.

"But it's worse for [a child]. And the reason it is worse for the children is that they are having the same experience as the soldier, but their brain is dynamically developing and those experiences are building into the architecture of the brain and it's changing them. It's building them in a different way. So, they are designed to expect threat . . . Their system is sort of always waiting, even if nothing's wrong . . . The vigilance is always there. And the reason the childhood piece is so important is because you're essentially building into the system of the person an incapacity to really relax and feel safe," Peters Bennett continued.

"And so the world continues to feel dangerous, and that's why so many of those children end up in juvenile justice. They end up in special courses in schools. They end up on the streets; they end up in foster care; they end up in prison—because they've never really been held in a state of safety."

After meeting with the Special Forces and considering their post-traumatic stress and, in particular, operator syndrome, Jamie could not stop thinking about the link between soldiers and prisoners. *Heroes and criminals. Warriors and prisoners. Hard to believe, but what drives one, drives the other,* he thought.

For him, it boiled down to this: If it was true that the majority of those in prison come from homes with toxic stress, creating heavy allostatic loads and eventually an overactivation of their sympathetic nervous system, then couldn't it also be true that recidivism rates would go down if the fight-or-flight mechanism was reset? If we took those criminals back to a pre-trauma state before releasing them back to the public sector, wouldn't that be better for everyone? If they could approach their environment clear eyed and mentally grounded, versus amped for danger and paranoid, wouldn't they be more apt to be successfully reintegrated back into a community?

INTENT AND THE BIOLOGICAL IMPERATIVE

Jamie found it interesting that mainstream media was starting to report on professional boxers and football players having severe violent impulse issues because of physical traumatic brain injury, finally allowing for discussion about science-based connections between brain health and criminal justice. But the conversations hadn't yet broadened to PTSI also being a source of violent impulses. Furthermore, it seemed that there was a biological imperative when it came to recidivism rates and impulse crimes that nobody was talking about.

In a 2022 interview, Jamie reflected on the true story of a friend who suffered a terrible concussion when he crashed a motorcycle and went from being gentle to criminally violent. "This happened a few years ago. Before the accident, my friend was a highly successful actor and a sweet guy, like any normal guy you'd come across."

The head injury, however, caused his friend's personality to change markedly, to the point that he was arrested on assault charges a number of times—though having never been in fights before the crash. A year and a half later, the man made national news for decapitating his caretaker's cat with his bare hands and then killing an old woman. Later that day, Jamie's friend fell to his death, most likely a suicide.

"It was all so shocking. I couldn't make sense of it. Not until I read Jay Faber's book on TBI and then started doing all this research on PTSI."

Jamie now believes much of impulsive violent crime can be linked to PTSI and TBI—combine these injuries and you have a person apt to get stuck in what Col. Toolan has started to call SOS (sympathetic overdrive syndrome), which is to have the overwhelming impulse to destroy yourself or others with immediacy.

"My friend's motorcycle accident caused head trauma, likely decreasing his executive function, but also likely tripping his sympathetic nervous system into overdrive when he was traumatized by the near-death experience. Then his allostatic load grew when he could not function mentally as he had before and his identity was stolen away. Sadly, for those with TBI and PTSI, being stuck in fight or flight is a lit match and the decreased executive function is kerosene. I don't care who you are, trauma biologically changes the brain and can lead to once nice guys decapitating cats. Or worse."

According to Jamie, Rae Carruth and Aaron Hernandez, two professional athletes who committed murder, had likely suffered from overactive sympathetic nervous systems as well as the obvious extreme traumatic brain injuries; they were dealing with paranoia and impulsivity and had very little executive function available to monitor their behavior and reactions. "CTE (chronic traumatic encephalopathy) and an overactive sympathetic nervous system is a deadly cocktail. But not one that needs to remain so."

Violent crime would likely dramatically decrease if brains damaged by concussion, disease, drugs, or alcohol were allowed to be repaired, which is exactly what Dr. Faber and Dr. Amen are doing. But post-traumatic stress injuries also need to be addressed for this to happen.

When an individual has an overactivated sympathetic nervous system, that biological change forces the amygdala to continuously fire off false signals of danger or stress, and that individual is not consciously able to control behavioral responses. The biological imperative built into our DNA, the behavior driven by instinct that is essential for an individual or species to survive, asserts itself, whether we want it to or not.

Logical behavior comes from a place in the brain that becomes secondary when the amygdala is firing off. That's the survival mechanism kicking in. It's just unfortunate that same mechanism kicks in when someone with post-traumatic stress is triggered by their kids yelling at each other about what cartoon to watch or a guy approaches on the sidewalk a little too close.

For those with complex brain trauma—like so many of the incarcerated—morality or desired intentions are not accessible when triggered. Their biological imperative is screaming, "You're about to be killed! You will have to kill in order to survive!" and so that is what they often do.

America is a nation of laws, and intent is the rule of thumb.

Consider this famous example: March 1993, Wilmington, Delaware. A key scene for gothic action-thriller movie *The Crow* was in production. The scene required actor Michael Massee's character to "shoot" star Brandon Lee with a .44 magnum Smith & Wesson model 629 revolver. Rushed and under pressure, the props crew had created dummy cartridges. Brass casings, tipped with real bullets, but without gunpowder, were being used—realism was called for because of the scene's close-ups. However, when the props crew loaded the revolver with blanks for the shooting, a bullet had separated from a dummy cartridge, lodging in the revolver's barrel. No one knew the bullet was there. While filming, the live round struck Lee in the abdomen, propelled by the harmless report of a blank—the errant bullet came to rest near Lee's spine, and he died hours later.

Massee did not intend to kill Lee, nor did the prop crew. The crew may have been culpable, or made bad decisions regarding procedure, but the perception was that they did not mean to harm Lee. Due to lack of intent, no one served jail time. If there had been evidence that the crew or Massee knew about the live bullet, or even if there had been a personal motive to harm Lee, then intent would have been established and someone would have been charged with a crime.

In that instance, there was no intent, so there was no crime. But when you broaden the idea of intent to consider the fallibility of a person with severe brain trauma who commits a crime, the issue is not so black-and-white. A person with severe TBI and PTSI is operating solely on the biological imperative level, and it's because of a physical injury creating repercussions often out of the control of the patient. The verdict would have been much different, however, if, say, Massee was someone quick to anger and paranoid due to TBI/PTSI and, in a moment of perceived threat, he shot Lee intentionally. If that had been the case, most juries back then would have found Massee guilty, unaware of the TBI/PTSI complications.

Jamie framed it this way: "If two guys with post-traumatic stress in the inner city get into an argument over a cheeseburger, and their sympathetic nervous system is telling their brain that conflict means life or death, and their uncontrollable impulse is kill or be killed, then how is the concept of intent applied when one guy ends up in the morgue and the other is in prison? Did they intend murder? Or was it an automated self-preservation response? Their stellate ganglion was lying to them.

"As a citizen of the United States, if our laws are based around intent and if neither of those men walked into that burger joint with the intent to hurt anyone . . . if both of their bodies had an automatic, biological reflex in response to a perceived threat to their life, then are they murderers or are they victims of a biological change that causes automatic, impulsive response? What do we do with people who we find to have a brain tumor in the prefrontal cortex after they've committed a terrible crime?

"Okay, the moral or philosophical interpretations can go on forever," said Jamie, "but the important thing to acknowledge here is that we bypass the argument entirely *if* we reset their brains to a pre-trauma state and they can respond with logic versus automated biological imperative responses.

"Because, sure, there are sadly always going to be assholes who actually like hurting other people, but I'm talking about giving the choice back to those who want it. Because, for sure, a good chunk of those who grew up in shitty neighborhoods with shitty parents and are stuck living out operator syndrome symptoms would rather have the choice to do no harm instead of just instinctively responding to phantom threats."

Certainly, society as a whole would be safer.

INSIDE THE PRISON

Fred Miles, a fifty-four-year-old inmate at Henrico County Jail, was incarcerated for possession of drugs and assault, in and out of (mostly in) prison for the last thirty years. Jamie approached him about being a part of the documentary on PTSI; Fred was eager to share his story.

"Why am I here?" Fred asked. "That's a loaded question. I have an issue with authority. I don't feel like I'm getting the help I'm requesting—that started when I was a child. Everybody that was supposed to be in a guardianship role over me was abusing me, from being a child on up. I learned to either fight or run."

Fred grew up in extreme poverty with no one to protect him, much like Trevor Beaman.

Trevor believed there was no difference in the trauma dealt to someone growing up on the streets from someone involved in combat overseas. In his own interview, the Green Beret said, "People are different, the places are different, but in my mind it's the exact same life, there is no difference. You can pull up West Side Chicago and drop it down in Kabul and it's the fucking same as Iraq. [In each of those places] you are so disconnected from the world, you don't have anything else. You ask yourself,

Do I leave the firebase and die today? Do I leave my house and I fucking get shot? The same trauma that's there, the same life, the same shit . . . it's the same exact feeling: I don't know if I'm going to make it home. Every day. And do that for twenty years . . . You want to know why people use drugs and alcohol and live wild and don't have any self control? It's because they don't know if they are going to live until tomorrow."

Soldiers were considered honorable, doing violence and killing others in the line of duty but always for the sake of their country. Those having grown up in the poverty and violence of inner cities, however, were not given the same pass, even when their mind was falsely telling them they were simply protecting themselves. Instead, they ended up in the prison system, like Fred.

Wearing a beige jumpsuit and leaning against a bone-colored cell wall in the penitentiary, Fred discussed his origins with Jamie.

"When I came out of my mother's womb, I was addicted to heroin," he said. "Me and my little sister, we was being watched by I think it was a friend of the family. He turned out to be a monster. You know what I'm sayin'? He molested me and my little sister and we were small. That was the first time I ever witnessed, at five years old, another human being killed. It's called *hood justice*. They blew his brains out right there in front of us. It's complicated to say where the trauma started, but from that point forward, for me, it was like, hell, what did I do to deserve this? I saw my sister brutalized, a grown man penetrating a four-year-old. I did what I could to protect her, but I was five. She still ain't right. I did whatever it took to get his attention so she could escape and go get help. What did we do as children to deserve that? Especially from someone who is supposed to be a guardian?"

The child rapist was executed in front of the two children, shot in the head at point-blank range. Raised by his older sister, Fred's father was a pimp, and his drug-addicted mother was killed by a john when he was still a child. Fred became a criminal and finally a murderer, unable to control his impulses. Within prison, his sense of hypervigilance and paranoia and need for violent response only increased in a population where he continued to live under a threat of danger. But he started to consciously process his past, and his responses, and his behavior, wanting to be better. He has studied philosophy and human behavior. He performs spoken word poetry as a tool for higher self-awareness, asking questions like: *What if life wasn't what it seems? What if God was hidden inside the shell of a crackhead or a dope fiend?*[41]

"I'm fifty-four years old," he said. "I've learned how to discipline myself, to control [my post-traumatic stress] a little bit better. You know, most young dudes around me with the syndrome—and they not even knowing they have it—they out of control. So, I'm like the voice of reason compared to them, but I still have moments and episodes where I can't control it, either. Like, I just got into it with the staff.

"They had to spray me down, put me in one of them chairs and chain me down. It shouldn't even have got to that point but it went from zero to a hundred like . . . you know, it was just out of control."

Fred acknowledged that he no longer wanted to react without pause, that he would like to be able to cut off his rage, to be able to trust himself. Since DSR is not yet an approved medical procedure for inmates in the United States, Fred planned on having Eugene's procedure as soon as he was released.

"Just imagine if everyone in the world was blind and then you could actually give them their sight and they could see, versus just what they've been experiencing their whole life, they could actually see colors for the first time. Or a person who has no taste buds and he can taste sugar or honey for the first time."

WHAT IF . . .

What could Fred have done with his life if just one door had been opened for him? Who can say? We can't know that someone would have made different choices, given the opportunity.

When Jamie first told Fred about DSR, Fred replied that if it is true that the shot does what it is purported to do, it would be revolutionary. "It would give me the opportunity to work and be well," he said. "I could be a human being again. Because I don't feel like a human being. To actually be able to just sit down, to eat a meal without being scared, no putting my back to the wall . . . I would love to be that person. It would be like a dream come true."

So what if violent tendencies could be turned down, or even turned off? Dr. Lipov's dual sympathetic reset has removed suicidal ideation fantasies from hundreds of reported patients, separate from the thousands of military folks . . . why not remove homicidal ideation? Do not both ideations come from the same physical injury, a place of darkness, lack of hope, the desire to remove that which pains you? The desire to be free from the tiger?

It was this kind of leap that led Jamie to Fred in the first place, not long after he had been introduced to the sheriff willing to allow access to prisoners who agreed to be interviewed about their trauma, and, with hope, be able to eventually receive DSR.

THE SHERIFF KNOWS

Jamie had first met Alisa Gregory at a consortium on brain health. He was surprised to discover she was a sheriff but also that she was there to gather information that might help with the care of her inmates.

As a matter of fact, Sheriff Gregory was the first woman *and* the first black person to hold the office of sheriff in Henrico County, a county incorporated more than four hundred years ago and now the fifth largest in Virginia. Furthermore, she was the only black sheriff in all of Virginia. She was an elected official, overseeing two large jail facilities, three courthouses, and civil process services in the American South, where her grandparents were born and raised under the crush of Jim Crow laws.

Jamie was impressed with her intelligence but also her empathy and conviction when it came to taking care of her "people"—people like Fred, who was serving time in her prison. He certainly wasn't the only one impressed; the sheriff prior to Gregory threw his weight behind her when he decided not to run for reelection. This, despite him being a staunch Republican who was in office for more than two decades and she a Democrat and a woman of color. Some of his constituents continue to be upset with him to this day. "They will stop in to the pizza place that he basically uses as his office and tell him off," said Jamie, after a visit to the town. "He's kind of got a Boss Hog thing going on, but he was brave to endorse Alisa Gregory for sheriff. She's a unicorn worth having in office, and he knew it."

Gregory grew up in Henrico County. She knows her community well.

"Some of these folks who wind up in here were classmates, or now children of classmates, and they were totally different people back then," she said. "I can flip through their mug shots and see the transformation, and yet not all of them have things happening in their life that would cause such changes. I asked myself, *What is that?*

"After all I've read in reference to brain health, and based on the studies and [what I've seen personally], I truly believe there are injuries to the brain that are causing some of the actions or decisions that these people are making that lead them down this road."

That revelation caused her to take a closer look at some of the individuals she'd once known as untroubled and yet who were now in her jail. "I found that a pattern was established: There was trauma. Unfortunately, a lot of men and women have experienced tremendous trauma in their childhood [but] they don't equate the actions they've taken or where they are now with the trauma that happened."

The sheriff had found the same PTSI and operator syndrome symptoms in her prison population that Jamie found in both the inner-city areas of extreme poverty, with his friend the businessman Michael Thomas, and within the military: high anxiety, hypervigilance, hair-trigger anger, being impulsive, not being able to sleep, a permanent sense of unease or impending doom. She agreed with Jamie that this syndrome existed everywhere, in every strata of society.

"I can really see these characteristics in multiple inmates," said Sheriff Gregory. "The lack of control, the anger, the feeling of hopelessness . . . but it's not just the [people] inside the jail. I think about the gun violence we are experiencing . . . all over.

People in a disagreement, they can just get so angry that they take another person's life."

Inside the prison, hypervigilance and paranoia were probably the most prevalent of these symptoms, she said. For good reason, as inmates were housed with others who would do them harm.

Henrico inmate Kevin Richmond witnessed another prisoner get his throat sliced open with the lid of a sardine can over a cigar. In a 2022 interview, he said, "I'm always on alert in these surroundings, on a daily basis. It could be nothing. It's just a part of me and everyday life . . . I feel good for a minute and things are going great, but then I'll hear something or see something and I'll remember shit's bad and think *I ain't gonna make it* . . . When I do sleep, it's very shallow, on my back, and alert. There is no sleep for me. Even if I'm at home, it's the same thing. I can't sleep."

Melissa McPherson had been to prison a dozen times on possession probation violation charges. She felt like there was "no good" in her and she hasn't experienced a day without unbearable anxiety. "I'm always waiting for the other shoe to drop," she said. Any emotion turned to anger and left her in fight mode.

Prisoner Cecelia Delacruz-Cenobio talked about wanting desperately to be free of the feeling of impending doom, and how she badly wanted to "just breathe." Even behind bars, she keeps a detailed schedule and checklist for everything, trying to get a handle on incessant anxiety.

The ever-present stress of growing up in deep poverty, or serving out combat missions, or the daily degradation and threat of violence in a prison creates the same symptoms, regardless of a person's position in society. Warrior or prisoner, it doesn't matter. When the combat is over, when the prison doors open, the individual will struggle to adapt to safe surroundings—and that struggle can be insurmountable if the sympathetic nervous system has permanently tripped the fight-or-flight response. The same can be true for an office worker or a stay-at-home parent.

THE DUTY OF AN ELECTED OFFICIAL

After four centuries of white, male Henrico sheriffs, Alisa Gregory won her seat in a landslide in 2019. She is now going on her twenty-fourth year with the county. She has a reputation for being tough and fair. Her mission statement revolves around the beliefs that she is an elected official who reports to her citizens and her actions are driven by their desires, and she is obliged to do what she can to rehabilitate those in her charge.

According to Gregory, there are 123 sheriffs in Virginia, eight of whom are female. "I think women bring a different perspective. We are naturally nurturers, and I think that's what we bring to this role . . . I come to work to ensure that not only my staff is taken care of, but that the inmates are also taken care of. That's always at the forefront of my mind when I make a decision, or when something's going on. I always think about, *If we do this, then how is it gonna affect staff? How is it gonna affect the inmates?*

"I look to make everybody's life better, because guess what. Everybody returns to the community eventually. The community that I live in. The community where I've raised my kids, where I'm helping to raise my grandkids. I have a vested interest in making sure that the people in this agency, whether they're staff or inmates, are happy. When they're in a good place and they go out in the community, then our community is happy."

She and her staff are sincere in their efforts to make a difference. The citizens of Henrico voted for a focus on rehabilitation rather than punishment. The judge, the prosecutors, and the jurors have heard the facts, determined guilt, and handed down a sentence, leaving the inmates accountable for their actions. "But when they come here, it is time to look at the person and be able to present them with resources . . . to help make them better than when they walked in here."

The stakes are high in Gregory's opinion. Simply warehousing inmates without interventions will allow whatever was going wrong in the prisoner's life to fester and become worse, and it can foster a culture of creating and strengthening criminality. This leads to a greater menace in regards to the welfare of the individual and the community in the future.

"Nobody is here because they want to be. Something in their life has gone wrong," she said. The sheriff encourages the inmates not to waste the time and resources they have available to them. They can change; they can find ways to be productive and grow. "And our staff can help, we have time to make connections or just listen. Sometimes we can get to the root of their problem in a way that nobody in the community may have been able to."

The Henrico inmate population fluctuates. Once hosting more than sixteen hundred prisoners, there are now under a thousand, thanks to the programs and hard work of the staff. An inmate has often been locked up because of bad choices, and sometimes bad luck, so the administration feels it is important to offer encouragement and hope, along with individualized release plans. They consider the pre-jail lifestyle and events, the criminal charges that landed them in jail, and their court dates. Medical and mental health issues are evaluated and treated in-house by the mental health staff, licensed clinicians, a physician, and a nursing staff. Classes or discussions on how to function successfully on a day-to-day basis in the outer world are important, like budgeting and balancing a checking account, how to do taxes, how to parent, how to fill out state forms or get a state identification number. This type of care and teaching has had an immediate impact on behavior on the inside and recidivism rates once they are released.

"There are some people we've actually been able to divert [out of the system] within a couple of days," Gregory said. The sheriff has managed to bring the inmate population down to what she believes is a safe number, in terms of being able to "serve my people." Gregory sees her inmates as a constituency she is there to serve rather than guard.

She would like to decrease the numbers even further, though, so she can have time to be working outside the prison walls. "If we had less inmates inside the jail, then we could spend more time in our community. We could be in the schools, greeting students at elementary schools and middle schools in the morning. It makes a difference. They get to see us in a different light. I talked to a young lady at a football game on Friday in reference to how she feels when there are police officers, firefighters, and deputy sheriffs shaking their hand when they come into school, and she was like, 'It's exciting.'

"With these programs, we create a relationship. Right now, there are a lot of children who aren't seeing police until things have gone bad. That is trauma. A traumatic incident. But going in and establishing a relationship with young people early means that they see us for what we really are: public servants. People who can help. So when those times come, when things have gone wrong, it's not a stranger coming to respond."

But for now, the staff at Henrico are doing what they can inside the prison system to get the help for those who need it.

GETTING THE HELP WE ALL NEED

Whether the changes in how the brain functions are due to TBI or PTSI or a combination, the resulting symptoms are often the same. Unfortunately, medical care for damaged brains and/or sympathetic nervous symptoms is a challenge to any doctor, considering that the issues are invisible to the naked eye, and that each individual brain reacts differently to brain trauma. Further, the SPECT scans are not generally available in the prison system, and definitely not the more expensive fMRI scans necessary to see the malfunctioning amygdala.

"I really hope more people start to listen and start to do the research needed to bring whatever treatment is going to help repair people's brains," says the sheriff. "And to make it available to everybody. You know, a lot of these people in the jail, they don't even have insurance, they can't take care of the basic medical needs."

The sheriff offers an array of treatments and medications to help keep her population healthy, including those with substance abuse and mental or emotional disorders. She has firsthand evidence that the common medical practices available to them are not enough when it comes to brain health. "Unless you get to the root of the problem and . . . then address the reason, they can't control those behaviors." She and her staff continue to see the same trauma-based patients over and over, even after they have gone through treatment after treatment.

Her ultimate goal is to provide a service that creates radical positive shifts in behavior before the inmates rejoin society, but she often feels like her prison officials are simply beating their heads against a wall. "I pray that more research can be done on trauma and brain health, and soon."

Back in California, Dr. Amen and Dr. Faber are doing just that, initiating treatment plans for those with TBI and/or PTSI using diagnostic tools that include a SPECT brain scan and a massive scan database. Like Eugene, they know that brain health is at the crux of a number of public health problems.

"A conservative approach to criminal justice is to scan [the brains of] the incarcerated and treat them, because then they won't go back [into society] and be a drain on governmental resources. They'll get a job; they'll pay taxes; they'll take care of their families," Amen said in a 2021 interview with Jamie and his team.

"It's not a political thing, not a left or right thing. It's the right thing to do. It saves them and it saves our society," said Amen.

If the "lock them up and throw away the key" people knew that most of these prisoners got out only to pose a far worse threat due to their trauma and increased criminalization from their time inside, they would think very differently, Jamie believes. Brain scans before and after would reveal the increased threat and the public health issue, and the need for DSR in prisons would be clear.

"The single most important thing I've learned is that mild traumatic brain injury can ruin people's lives. And nobody knows about it because nobody is looking at their brains," says Daniel Amen. The majority of providers in the mental health care industry are not nearly as effective as they could be, according to Amen, because they are not visually assessing the organ they are evaluating. When they do not utilize brain scans, doctors are unnecessarily choosing to not use available neurotechnology to diagnose a patient's issues. "No helicopter pilot would fly blind. No orthopedic doctor would fly blind . . . Diagnostic methods have barely progressed since the days of Abe Lincoln."

Dr. Amen trained as an X-ray technician in the army during the Vietnam era. He left the armed forces to go to medical school but then returned as an active-duty army officer and completed a psychiatric residency at the Walter Reed Army Medical Center. Now a double-board-certified psychiatrist, he was intimately aware of the mental health issues found in the military—but as a neuroscientist and civilian trauma physician, he recognized brain health as a societal issue.

Sheriff Alisa Gregory and Dr. Daniel Amen are right. The medical industry needs to treat brain health across the board, in all people who are suffering from TBI and/or PTSI.

And that includes prisoners. SPECT scans have been done on drug-addicted criminals, a few of which have now been compared to a post-war combat veteran scan—revealing the same lack of blood flow in the same areas and evidence of trauma in the brain, according to Dr. Faber.

Doctors acknowledge there is a severe PTSI problem in the military, affecting soldiers, their families, and the larger population, so why not acknowledge that the same trauma issues in prisoners are harming the individual but also society in general?

Dr. Amen performed a study with the medical director at Sierra Tucson, a large psychiatric and addiction treatment facility in the Arizona desert. Out of five hundred consecutive new patients, 44 percent of them had a significant history of brain injury, which aligns with what he has seen at Amen Clinics.

"If eighty to ninety percent of people who are incarcerated have been or are addicted to drugs or alcohol," says Amen, "that statistically means they have toxified, troubled brains. Of that, at least half of them have had a traumatic brain injury in their past that was significant."

STICKING OUR HEADS IN THE SAND

Sheriff Gregory is fascinated by the science behind taking an unhealthy brain and making it healthy again. She began researching different therapies and nutrition and supplements, searching for anything that might benefit those in her facility, which is why she was at the brain health symposium with Jamie in the first place. "I'm looking for anything that will help these people. Right now, what we know and continue to do isn't creating the results we want," she said.

Many prisoners have undergone deep trauma leading to PTSI, but thanks to Amen, with supplements and sobriety, TBI caused by drugs and alcohol can be reversed, giving a person a biologically fair chance, and with Eugene's DSR, they can be brought back to a pre-trauma state . . . meaning significantly less mental anxiety, fewer physical ailments, less criminal behavior.

The injections have not yet been approved by the federal government to be used on inmates, so Sheriff Gregory continues to do what she can.

Sara Harman is the Henrico County programs director and personal counsel to the sheriff after being in private practice as a criminal defense attorney for fifteen years. "I've defended some people who have done horrible things," she said. "I see people here who have done some horrible things; they've made terrible decisions. But every person that I come in contact with has a redeeming quality. There's some good to that person, always.

"So, when you see people . . . doing bad over and over and over and over again, there's gotta be an explanation. We're missing it; we're missing something. We need to start by figuring out what's going on in these people's brains. I think that's where we start because, without it, we're just addressing things like lack of education and, you know, the poverty mandates, which just put Band-Aids on things.

"I've seen things that have worked, things that we've been able to do in the jail that have worked, and people have ended up never coming back here. So, there is some success with rehabilitation and reentry into the community. But I think we're still missing a big part. The men and women prisoners, they're all getting out of here and they're gonna be a part of our communities. If you stick your head in the sand about it, you are part of the problem."

Dr. Faber, who uses brain health care to help troubled youth stay out of prison, agreed. "Should someone with a bad brain who hasn't been treated leave prison? Could they not in fact be a danger to other people? That could lead to potential violence or destruction . . . if it's not properly addressed? Well. Yeah. I mean, it's kind of common sense. If [a prisoner has a sick brain], shouldn't we begin to start addressing and treating it before they are released?"

REFRAMING THE SYSTEM

As was discussed earlier, the impulse to commit crime is a biological imperative for those who feel threatened and have impaired brain function due to trauma. The science backs the necessity to reframe the criminal justice system. Ignoring brain health issues, including a hyperactive sympathetic nervous system, has created a public health issue—including letting people out of jail *knowing* they are going to commit crime again.

Clear back in 2012, *Scientific American* reported, "In prisons . . . approximately 60 percent of adults have had at least one TBI—and even higher prevalence has been reported in some systems. These injuries, which can alter behavior, emotion and impulse control, can keep prisoners behind bars longer and increase the odds they will end up there again. Although the majority of people who suffer a TBI will not end up in the criminal justice system, each one who does costs states an average of $29,000 a year."[42]

Trauma can change the architecture of a child's brain, but it can also take away choice from an adult. If your brain is operating under the assumption that you have been threatened, your body will react, whether you will it to or not.

Any intervention that brings the nervous system down into a state of greater calm and tolerance is key. If those with PTSI can manage their reactivity—if they can consciously acknowledge when they are triggered and can do things to regulate their body and mind's response—then our world is going to look a whole lot different. We might just be able to stop those with PTSI from committing impulse crimes, possibly even correct their mental injuries as traumatized youth before they can act on the impulses.

"I think now we are at a place where we really can change the narrative and focus on the fact that it is truly a public health issue," said Sheriff Gregory, discussing solutions with Jamie on camera. "There is talk about the lack of being able to address the mental health issues, the lack of being able to get control and address the substance or the abuse disorders. You can throw money into it, you can build more state hospitals, you can build more recovery centers, you can develop more medications to be able to suppress the desire to do drugs, but until you get to the heart of the problem, what's really going on with each person—in their brain—those problems are gonna resurface."

Again and again, the inmates who spoke with Jamie reiterated a desire "to just be normal."

Regardless of one's thoughts or stance on criminals and justice in general, isn't it in the best interest of the community that a prisoner, when released, is no longer operating under criminal impulses? If recidivism rates drop, that's good for all of us.

"I'm not talking pie in the sky here," says Jamie, "I don't think society all of a sudden embraces an ex-con because he's had a shot that 'calms his nerves.' Obviously, transitioning from a prison culture to a city neighborhood with all its politics and stigmas is going to remain difficult for those individuals. I'm just saying if he's not in fight-or-flight mode all the time, he's less likely to stab someone for getting pushy waiting in line at McDonald's. He's less likely to carjack you in your neighborhood because he's not stuck in kill-or-be-killed mode. He's able to stop and process the situation versus reacting impulsively, on a simple survival level."

Jamie believes shifting our perception to view impulse crime as a major public health issue goes a long way toward a safer society.

Frankly, the injury caused by post-traumatic stress is the cause of many, *many* public health issues, some of which are addressed in the next chapter—and all of which can be healed at least to some degree with treatments like DSR becoming more readily available.

CHAPTER SEVEN

The Crisis

As individuals, we now spend hours each week at doctor visits and picking up prescriptions. Some of us spend that time picking up a bottle at the liquor store or turning to another vice. We desperately want our sickness, our pain, our anxiety, our restlessness, our sadness to go away. Many do not want to feel at all. But what if we could get to the root of what is causing this malaise? What if we could get to the root of the illnesses that are keeping us from work and family?

"The results of my study from 2022 showed that in the twenty-two different types of trauma that we've seen, everybody responded to DSR. Trauma is trauma," said Eugene.[43] "People are miserable."

It is time to dig deep and assess the breadth of the public health crisis. A good chunk of the human race is living as if that tiger is just around the corner, ready to pounce—and that unrelenting stress is wreaking havoc on the body as much as the mind. And the cost of a large population of sick people harms society as much as the individual, creating a burden on the country's health system.

Data science and the top doctors in multiple fields support this view of post-traumatic stress and the medical consequences; the problem is that the general public remains largely unaware of the proliferation or the bigger picture of the threat they are dealing with: every person on this planet will suffer trauma to some degree, and if it's bad or long term enough, no matter your job or your class, your trauma is going to show up as post-traumatic stress, which will eventually lead to heart attacks, cancer, or other illnesses.

In a CDC study published in 2002, the relation between post-traumatic stress and perceived physical health was investigated. While these same outcomes apply to everyone across every level of our society, this study was done by the military. "Participants included 3,682 Gulf War veterans and control subjects of the same era who completed a telephone survey about their health status. Veterans screening positive for PTSD via the PTSD Checklist–Military Version reported significantly more physical health symptoms and medical conditions than did veterans without PTSD. They were also more likely to rate their health status as fair or poor and to report lower levels of health-related quality of life. The results of this study are consistent with studies of other combat veterans and provide further support for an association between PTSD and adverse physical health outcomes."[44]

It is unfortunate that the lion's share of data comes from the military, including case studies and interventions, if only because it makes it appear to be most prevalent in that population, but the larger world is starting to wake up to the realities.

According to Eugene, "In the real world, this public health crisis looks like issues such as absenteeism . . . a lot of people are not available because they are dealing with the struggle to simply get out of bed. Or it looks like alcohol abuse. Or like two to three times higher rates of heart attack. It looks like cancer, or rheumatoid arthritis, or autoimmune diseases.

"Post-traumatic stress causes a domino effect of physical illnesses to happen within the body. There is no denying it. This is fact."

Why? It's about brain chemistry. When the overactivation to the sympathetic nervous system occurs and the fight-or-flight mechanism is stuck on, the brain's neurotransmitters become imbalanced. To break it down, *Medical News Today* states, "The nervous system controls the body's organs, psychological functions, and physical functions. Nerve cells, also known as neurons, and their neurotransmitters play important roles in this system. Nerve cells fire nerve impulses. They do this by releasing neurotransmitters, which are chemicals that carry signals to other cells."[45]

"In other words," says Eugene, "post-traumatic stress can lead to autoimmune dysregulation, which is demonstrable by blood tests. Having PTSI means the autoimmune system will not work correctly. For instance, if you don't have a competent immune system, then it cannot do scavenging for cancer cells. If a cancer cell has been produced, it likely will be chewed up by the immune system's 'killer' scavenger cells,[46] and the person doesn't get cancer. But if someone is on immunosuppressants, like heart transplant patients, the chance of developing cancer is about five times higher because the scavenging process is not working."

Other potentially dangerous health issues that can occur if the neurotransmitters remain out of balance:

- Adrenal fatigue
- Hormone imbalance, which creates thyroid problems, low testosterone, polycystic ovary syndrome, and obesity
- Chronic fatigue syndrome
- Digestion-related dysfunction and disease
- Dermatologic issues

The list goes on. And on. The following pages outline just a few of the biggest, most documented issues related to PTSI, focusing on the effects of post-traumatic stress first at the personal level and then at a broader societal level.

PTSI AND CANCER

"I have the perspective that if you can fix the immune system, bringing it back to a normal state, I think the chance of cancer will be reduced," says Eugene. A research study done in Ireland compared women who underwent mastectomies using general anesthesia versus T2/T3 ganglion blocks. The Irish doctors followed up with the patients five years later and found that there was a marked reduction of returning cancer, 30 percent less than expected, in the women who had the T2 blocks. Eugene believes they were able to fix the autoimmune response by resetting the sympathetic nervous system using that block, which in turn balances the brain's neurotransmitters.

The American Psychological Association agrees with this proposition, saying in the *APA Working Group Report on Stress and Health Disparities*, "New models and research are helping to identify the mechanisms through which socioenvironmental factors, including experiences of stress, influence cancer progression. Under normal conditions, natural killer cells help to prevent the spread of tumors. Under chronic stress, for some cancers, stress hormones suppress the activity of natural killer cells and disrupt a cascade of processes, which inhibits the destruction of tumor cells and other processes."[47]

In New York, there was a study on women who'd undergone mastectomies that showed that 25 percent of them came away from the surgery with post-traumatic stress symptoms. "Those PTSI symptoms suppress the immune system, so if that is allowed to continue, the cancer will come back in those women," says Eugene.

Another study published in June 2011 looks at how stress has been implicated in the development of cancers and confirms the role of adrenaline (levels of adrenaline increase in response to stress) in colon cancer.[48]

There are many personal stories to support the statistics linking cancer to PTSI, including that of filmmaker Corey Drayton. Corey believes the repression of shame and abuse that occurred throughout his upper-middle-class childhood left him dealing with a growing allostatic load, which lead to PTSI and a body primed for chronic illness—including, finally, stage IV colorectal cancer, which spread to the prostate. In a 2022 interview, he stated, "The initial traumatic conditions leading to my eventual cancer began with the deep-seated alienation I experienced from birth. I was born sickly and premature. Needing to be incubated, I had zero human touch or contact for the first month of my life. Being the accidental, only child of two workaholics who had no interest in being married—let alone being parents—I continued to grow up lonely and isolated.

"Luckily, my family was affluent, but . . . my position in the family was as an accessory; my job was to attain as much self-sufficiency as possible and to not be a bother to my parents. We moved constantly, so I had little continuity with other kids and, in the eighties, one had to put in real effort to maintain a friendship over great distances. The geographical distance wasn't the only distance I experienced, however: There was the distance of heritage, being biracial black/white and Jewish I didn't belong anywhere. There was the distance my natural introversion created between myself and others. There was the distance created by my parents' tumultuous marriage. Every year they orbited the black hole of divorce."

Corey's childhood health also played a role. "I was a slight child with a modest constitution who suffered severe asthma. I experienced my first asthma attack at age eight, hallucinating and losing consciousness . . . I was being hit at home, verbally abused. I developed a blistering hatred and disgust that aimed inward like some Death Star super laser."

He first attempted suicide at age eighteen. In his teens and then again in his early twenties, he'd been forced into unwanted sexual acts, accentuating his self-loathing and struggle to establish healthy relationships. He cycled through multiple abusive relationships with narcissistic, cold women. His work relationships in the film industry were often no better, as he found himself in a "woke" work environment that was blindly yet overtly racist.

After a few serious health scares (which healthcare providers minimized), he was finally diagnosed with cancer at age thirty-six. "It was at stage IV. It had metastasized, invading my prostate and my lymph nodes. I was given a twenty-seven percent chance of survival. A course of radiation, chemo, and eventually colostomy and prostatectomy surgeries were planned. On the last night of Hanukkah, I had a massive hemorrhage, pouring out blood on the bathroom floor of our flat. I lost consciousness (my third near-death experience) and was taken to the ER. It took four blood transfusions to stabilize me. I spent a month in the hospital. My mother had to argue with my father to get him to come see me. Over the next two years, I weathered the slow grind of chemo and daily radiation therapy. My partner at the time left me after the diagnosis, adding an urgency to my feelings of isolation and abandonment.

"I think the cancer largely manifested out of an internalized existential doubt: Was I wanted? Where did I belong? Who was my tribe? Was my body even my own? The lack of human touch in the crucial first weeks of my life was really at the root of it. As a species, humans need that ineffable maternal reassurance that exists in the neural space of skin-to-skin contact, to hear the thrum of your mother's heartbeat, to feel her warmth. I didn't have that."

Corey survived the cancer and the treatments. But it wasn't until he made the decision to focus on the what, why, and how behind his physical issues, and made strides toward mental and emotional well-being, that he felt like he was beginning to gain control over his overall health.

"I had DSR in autumn of 2021. It worked. I felt an immediate rush of all of the emotions and somatics that I hadn't felt in decades. Both sides resulted in an intense discharge. I woke up sobbing and also recall feeling a lack of inhibition; I had an openness to connection and exchange with complete strangers that I had never experienced. Cancer can be extremely isolating. I think cancer survivors should ask themselves: Do I have traumatic stress? Is it possible that I am also dealing with something that cannot be healed by chemotherapy, radiation therapy, or surgeries?"

HEART DISEASE

A common refrain on daytime doctor shows centers around how stress can lead to heart attacks. There have been dozens of studies showing that heart disease and heart attacks are twice as likely when somebody has PTSI. A government report from 2017 states, "Though studies have found patients with psychological trauma and PTSD are at greater risk of a variety of chronic physical ailments, associations with cardiovascular disease (CVD) are particularly concerning."[49]

"As always, the question is . . . why is that?" asks Eugene. The response is becoming repetitive: When someone has PTSI, the fight-or-flight mechanism is overactive. That overactivity will cause a continual and chronic "squeeze" on the blood vessels. Over time, the blood vessels become smaller, leading to plaque in the coronaries. Which leads to a heart attack. In PTSI patients, that escalates the danger of death.

"Numerous population-based studies have demonstrated that patients with PTSD are more likely to develop and die from CVD. These findings have been confirmed in diverse populations, including Veterans and active-duty military personnel, nurses, and 9/11 survivors," according to the National Center for PTSD.[50]

ERECTILE DYSFUNCTION

Like many of the men who come to Dr. Lipov for treatment, John Smith (a pseudonym) didn't mention his sexual dysfunction (SD) at first. As a combat veteran who'd fought in the Vietnam War and who had gone on to build a career in law enforcement, he lived with a diagnosis of complex PTSD, and his sex life wasn't his top priority. One psychiatrist who diagnosed Smith told him he had the most severe case of PTSD the doctor had seen in thirty-five years. During a medical research study he participated in, a researcher ran into the room wearing a hideous mask and a noisemaker, screaming and yelling. "I flipped the table over, ran at him, jumped on him and beat him to the floor," Smith recalls. "They pulled me off. Apparently, I was the only one who had the 'fight' response."

Smith ended up getting the DSR three times because of a neck injury that made it harder for the anesthetic in the SGB to have its full effect. One symptom that unexpectedly went away was his sexual dysfunction. As his wife told Eugene during a follow-up visit, his equipment was now working just fine.

"I am not Masters and Johnson," says Eugene, "and I don't take patients' sexual history. But based on published reports, eighty to eighty-five percent of men with PTSI have sexual dysfunction. It makes sense that sexual dysfunction would accompany the fight-or-flight response. If you're running away from danger, that's not the right time to reproduce. Your norepinephrine levels go way high. Without this hormone at its optimal levels, you will have no inclination toward or ability to have sex.

"And it isn't only men who suffer from sexual dysfunction associated with PTSI. When women are under stress, their progesterone levels drop. Progesterone is needed for sexual desire."

Sexual dysfunction, severe enough to interfere with quality of life, has been reported in a large number of veterans clinically diagnosed with PTSI. A number of causes have been discussed, including overactivation of the sympathetic nervous system.

To which, Eugene says, "Duuuh."

A study by Dr. Kenneth Hirsch reported that over 80 percent of the study's combat veteran patients "were experiencing clinically relevant sexual difficulties."[51] In a study of Vietnam veterans, another researcher demonstrated that the three catecholamines—dopamine, norepinephrine, and epinephrine—were all excreted at an elevated rate in patients with severe PTSI, and that levels of these three catecholamines seem to be particularly related to intrusive symptoms, which may contribute to the phenomenon reported by patients who experienced intrusive images during sexual relations.[52]

Available evidence also suggests that autonomic arousal, anger/hostility, and relationship difficulties—characteristics of post-traumatic stress—may contribute to the observed sexual dysfunction. The anger regulation deficits have been demonstrated in combat-related post-traumatic stress injury.[53]

Can normal sexual function be a natural antidote to this trauma? Can lack of normal sexual function contribute to persistent trauma injury in men? Much is left to be discovered in the association of post-traumatic stress and sexual dysfunction, though it seems likely that sympathetic modulation will be an important tool to improve sexual function, and SGB may contribute significantly to the treatment of SD.

COVID-19

Post-traumatic stress does not cause COVID (though it does weaken the immune system, making it easier to contract the virus and harder to get rid of it), but in some patients, COVID does cause post-traumatic stress—COVID infection can lead to sympathetic nerve overactivation.[54]

There is both a physical and an emotional trauma that occurs when someone develops a severe case of this disease. As of December 2021, 276 million people around the world have been diagnosed with the coronavirus called SARS-CoV-2, and 5.3 million have died, according to the CDC. In America, 51.4 million have been sick with the virus, and the death toll is over one million. In relation to post-traumatic stress, Eugene says, "The virus increases the chance of an overactive sympathetic system. We know COVID increases the chances of suffering from PTSI because, for one, people are traumatized by the looming threat of death and dealing with long-term isolation and all that, but it also physically activates the fight-or-flight system, again, increasing the chance of PTSI."

Of further interest here is the research being done on using DSR with the large subset of those who develop a range of symptoms that persists for many months, what the public calls *long COVID*. The constellation of symptoms, known formally as PASC (post-acute COVID-19 syndrome), can include fatigue, shortness of breath, brain fog, sleep disorders, fevers, gastrointestinal symptoms, anxiety, loss of taste, change in taste, loss of smell, and others.

Eugene has been working with an anesthesiologist from Alaska, Dr. Luke D. Liu, who has published findings about how his use of the stellate ganglion block reversed long-COVID symptoms—after the injections, the patients regained their sense of taste and smell.

One of the patients came into Dr. Liu's clinic after suffering extreme symptoms for seven months. "She continued to experience debilitating fatigue and speech impediment, and had returned only to limited duties and shortened hours at work. She reported that ongoing dysgeusia [a disorder that alters or impairs sense of taste] had led to food aversion and significant unintentional weight loss. She was experiencing severe generalized body pain." But at the sixty-day follow-up appointment after undergoing DSR, "she reported normal levels of fatigue and cognitive function, durable restoration of smell and taste, and absence of post-exertional malaise."[55]

According to Eugene, "Since the report, Dr. Liu has successfully reversed long COVID symptoms with about forty people using DSR. So, the block seems to work against PTSI and post-COVID symptoms. Isn't that fascinating?"

The psychological destructiveness of this pandemic on healthcare workers and first responders has become increasingly important. "The Coronavirus Disease-19 (COVID-19) pandemic has highlighted the critical need to focus on its impact on the mental health of Healthcare Workers (HCWs) involved in the response to this emergency. It has been consistently shown that a high proportion of HCWs is at greater risk for developing

Post-traumatic Stress Disorder (PTSD) and Post-traumatic Stress Symptoms (PTSS),"[56] according to an October 2020 report from the Public Health Emergency Covid-19 Initiative.

"The DSM-5 indicates that 'experiencing repeated or extreme exposure to aversive details of the traumatic event(s)' can be considered as potentially traumatic events (criterion A4: e.g., first responders collecting human remains, police officers repeatedly exposed to details of child abuse).

"Healthcare workers in emergency care settings are particularly at risk for PTSD because of the highly stressful work-related situations they are exposed to, that include: management of critical medical situations, caring for severely traumatized people, frequent witnessing of death and trauma, operating in crowded settings, interrupted circadian rhythms due to shift work."

Dr. Robbyn Peters Bennett has had a number of patients dealing with COVID fallout. "The allostatic load is an important thing to think about with the pandemic. With the COVID situation, people stress and stress, and over time you get worn out. You're not getting enough recovery and you're not getting that opportunity to be in deep relationships or to ease up and just enjoy and to feel safe and good or loved. Right now, there's too much threat impinging on us. And if that happens consistently over time, that creates an allostatic load and it makes your body vulnerable to all sorts of mental and physical health problems."

Caregivers and teachers are two example civilian areas that have been hit particularly hard with the physical, emotional, and mental strain of dealing with the virus itself, as their jobs require proximity to those who are likely ill, but also the massive upheaval around adhering to new policies and practices (i.e., making others wear masks and socially distance) that have become so fraught with politics and rage. Those working in the restaurant service industry are another example.

Hopefully, if you are working in an environment such as this, you are reaching out to the medical profession for support and care. Keep in mind, COVID may seem like a momentary blip you simply have to see through, but the trauma around the illness and the societal response can sit in your body for a long, long time, wreaking physical and emotional havoc.

ADDICTION

Why do people use drugs? Why do they use alcohol? "They are trying to take their troubles for a swim," says Eugene. "Let's say it's alcohol. They keep drinking; they are trying to get to a point where they are not sensate. They are not feeling the pain. They're trying to find something to alter their consciousness."

According to the American Addiction Centers, "Those diagnosed with post-traumatic stress disorder are three times more likely to abuse substances . . . People seeking treatment for PTSD are fourteen times more likely to also be diagnosed with a substance abuse disorder (SUD) . . . Research has found that service members and veterans who have heavy drinking tendencies are more likely to have PTSD, depression. War veterans with a PTSD diagnosis, who also drink alcohol, tend to be diagnosed with binge drinking."[57]

Renowned Hungarian-Canadian physician and bestselling author of, among others, *The Myth of Normal* and *In the Realm of Hungry Ghosts,* Dr. Gabor Maté has spent decades working with the impact of trauma on our biology, often with a focus on addiction.

"When you look at the literature on what causes stress for people, it's uncertainty, lack of information, loss of control, and conflict. Now, if I had to design a society that's gonna impose those stresses on a large population, I'd design exactly the society that we have right now," says the doctor in a 2022 interview with Jamie.

Dr. Maté also defined addiction as "manifested in any behavior in which a person finds temporary relief or pleasure and therefore craves it, but then suffers negative consequences in the long term and does not give up [the addictive behavior] despite the negative consequences . . . that can include drugs, but it could also be things like shopping, eating, work, gambling, sex, relationships, and pornography . . . So, that relief that people feel [once they've displayed the addictive behavior], that means that the primary problem is not the addiction. The addiction is an attempt to solve a problem: the problem of discomfort with the self, the problem of lack of peace inside, the problem of emotional pain, the problem of stress, the problem of alienation. Addiction always comes along as a solution, as an attempted—doomed, but an attempted—solution to a problem. Now the question is, where did that problem come from? Not from genes, and not from a choice, but from trauma. My mantra on addiction is not, "why the addiction" but "why the *pain*?" The pain comes first. The addiction comes second. And that pain always originates in trauma—always, always, always, always."

Dr. Maté believes stress affects us even before we are born. "[Our] biology is affected by life experience and emotional factors, beginning in uterus. You can stress pregnant mothers, whether animal or human, and that'll affect the biology of the brains of their yet unborn children in a way that'll predispose [the child] toward addiction. And so on through the life span so that, yes, there's a biological dynamic inseparable from the psychological circumstances."

He also discusses how the lack of parental bonding, which establishes normal brain chemistry and initiates natural endorphin highs in infants and children, can be a cause of later addiction. The doctor believes those who don't experience the early endorphin rush that comes with feeling loved will often seek out replacements, like the rush experienced with opiates or alcohol.

"So, who's to be blamed for a childhood in which the circumstances did not favor the development of the appropriate brain circuitry? People can't be blamed for their biology. Their biology develops as it develops, based on circumstances, because we know that the human brain develops an interaction with the environment. We're not just talking about psychological factors that induce pain, we're also talking about psychological factors that affect the very biology of the brain, which then later on leads to certain behaviors. So, the biology is inseparable from the psychological experience, and trauma interferes with the biology of the brain . . . This is why we have addiction."

At the Stella Center, Dr. Lipov and the other clinicians have treated over five thousand post-traumatic stress patients. In a 2013 report, he found that "there are numerous adverse consequences of comorbid PTSI and alcohol use disorder, the most alarming of which is that mortality is more than twice as high among patients with PTSI and comorbid alcohol dependence as compared to patients with PTSI alone."[58]

But he has also seen the other side. He says, "We've had a lot of people who stopped drinking alcohol. There was a lieutenant in the Special Forces; he was drinking a liter per day of hard liquor.

We did the SGB and he stopped drinking. Completely. His mind cleared, he claimed to have no withdrawals. It seems to work with those with PTSI who have become addicted to alcohol and smoking while trying to suppress their angst, anger, fear, or flashbacks."

While still in the early stages of understanding how DSR worked on hot flashes, Eugene found an article in *Nature* that talked about how people who'd had a stroke in the insular cortex region of the brain, next to the hippocampus, discovered upon recovery that they no longer had a desire to smoke. The insular cortex is known to control all addictions, and Eugene knew there was a connection from stellate ganglion to the insular cortex. Eugene was amazed at the time but now has seen it multiple times firsthand; he recently performed the SGB on a woman who had been smoking for over sixty years—after the procedure, she was able to quit cold turkey.

He then treated fifty other patients addicted to smoking who were able to stop smoking after DSR, and he now believes that there is the potential that DSR can work with most addictions, including sex addictions or gambling addictions.

Substance abuse affects more than just individuals. Those with addiction issues cause harm to their families and the communities that must care for them, physically and financially. Dr. Faber has focused much of his brain health research in this area. According to Faber, addictive cravings are reduced significantly when the brain is rehabilitated. He says, "Eighty-five percent of people who are currently incarcerated have some type of substance abuse problem. Eighty-five percent! Okay, what does that mean? It means that if we could take those eighty-five percent and help them, we could see our prison population drop to fifteen percent and the amount of money we'd be saving per year per capita would be huge."

Dr. Maté says, "The significance of our capacity to reset the brain and to reset the nervous system [with DSR] is that, in doing so, people would not have to pursue their ingrained behaviors and would not be stuck in their traumatic imprints . . . If we can provide environmental conditions in which the brain can reset itself, in which the autonomic nervous system can be reset . . . then that's where healing happens . . . The significance of that would have to be recognized by the legal system and by the medical profession, which would be revolutionary."

SUICIDE

While the physical ailments caused by post-traumatic stress can pile up, suicide is far too often the ultimate dark outcome if that trauma injury is left untreated.

The all-too-real statistics around suicide linked with post-traumatic stress are staggering:

A report by Cambridge University Press in July 2009 found that individuals with PTSI were approximately fifteen times more likely to attempt suicide than those without PTSI, and the association remained after adjusting for depressive symptoms.[59] According to a 2021 Swedish study, the risk of suicide is particularly high for women with PTSI, primarily attenuated by suicide attempts before the PTSI diagnosis. The study covered 3.1 million people, a crosshatch from all populations, and found that individuals diagnosed with PTSI are twice as likely to die by suicide than those without PTSI.[60]

"Post-traumatic stress disorder is an independent predictor of attempted suicide. Exposure to traumatic events without PTSD is not associated with a later suicide attempt." This from a report published in June of 2018, following a randomized trial of American young adults that became a cohort study. A total of 1,698 young adults (mean age, twenty-one; 47 percent male; 71 percent African American) represented 75 percent of the original cohort of 2,311 persons, with the following results:

"Post-traumatic stress disorder was associated with an increased risk of a subsequent suicide attempt. The PTSD–suicide attempt association was robust, even after adjustment for a prior major depressive episode, alcohol abuse or dependence, and drug abuse or dependence (adjusted relative risk, 2.7; 95 percent confidence interval, 1.3–5.5; $P < .01$). In contrast, exposure to traumatic events without PTSD was not associated with an increased risk of attempted suicide."[61]

The bulk of statistics collected around suicide and PTSI are still focused on the military. Recent wars have led to a large number of military personnel having multiple and severe symptoms of psychological distress, and this increase has been punctuated by reports of extremely high suicide rates. Though this population of PTSI sufferers are the most studied, barriers to their care continue to exist among military personnel; namely, the stigma attached to having a mental health issue and the difficulty finding a treatment regimen that works.

From the invasion of Afghanistan in 2002 until summer of 2010, the US military lost 761 soldiers in combat—yet, astoundingly, we lost 817 soldiers to suicide during that same period. A recent document filed in the Ninth Federal Circuit Court of Appeals claimed eighteen veterans per day take their lives and one in four of them is enrolled in the VA medical system. Among all veterans in the VA system, one thousand attempt suicide each month.

The numbers can be overwhelming. But they do not need to continue to grow.

At an APA presentation in 2015, Dr. Lipov spoke about his use of SGB (pre-DSR) in treating suicidal ideation: he had found that patients with impulsivity and sleep disorders, especially nightmares, have an increased risk of suicidal ideation and suicidal activity.

"But what does DSR work for? That's right: impulsivity and sleep disorder," says Eugene. The first person to publish about the reversal of suicidal ideation in relation to SGB was Dr. Justin Alino in 2013. A patient of Dr. Alino's had suffered significant abuse as a child, became an alcoholic as a teen, and served two combat tours, where he reported multiple convoy attacks and firefights and seeing burning or dismembered bodies. The patient was diagnosed with PTSD. He tried to kill himself three times. He underwent the stellate ganglion block, and in two days his PCL went from 85 down to 18, which is normal, leaving him clear of suicidal thoughts.[62]

"This is because the block corrected the invisible machine. You reduce impulsivity and sleep disorder in PTSI patients, you can take away suicidal thoughts," says Eugene. In an abstract he wrote in 2015, "The Use of Stellate Ganglion Block in the Treatment of Panic/Anxiety Symptoms (Including Suicidal Ideation), with Combat-Related Post-traumatic Stress Disorder," he provides operational evidence of suicidal thoughts tapering off after the procedure, with a focus on one veteran in particular.[63] His story of previous treatment history, severity of symptoms, and the need to find a clinician outside the conventional military providers demonstrates the urgent need for a system-wide prevention change.

The veteran presented with severe PTSI and active suicidal ideations. Eugene treated him with the stellate ganglion block and had resolution of the suicidal ideation within thirty minutes following the procedure. The pain physician repeated the treatment one more time, sixteen days after the first. Four years after completing the treatment, the patient was reassessed via psychiatric measures to show a very marked reduction in symptoms, and he self-reported that he'd had no further suicidal ideation since the injections.

If a person believes themselves to be under the threat of death twenty-four hours a day, every day, then eventually they will no longer want to live. But when the amygdala stops sending distress signals, then the suicidal ideation can stop.

Green Beret Trevor Beaman is the perfect example of this, having amassed horrific trauma in his childhood, only to have it grow with his time in combat zones. He attempted to take his own life multiple times, despite years of expensive treatments he received via the military. It was no wonder he was skeptical of DSR, but to this day he continues to be free of suicidal or homicidal ideation.

ADVERSE CHILDHOOD EXPERIENCES (ACES)

Adverse childhood experiences as a precursor to PTSI and a plethora of other illnesses was introduced in earlier chapters, in the stories of Trevor and prisoner Fred Miles.

There are many individuals impacted by ACEs, enough that the societal consequences are massive.

According to the Centers for Disease Control and Prevention, "Adverse childhood experiences (ACEs) include verbal, physical, or sexual abuse, as well as family dysfunction (e.g., an incarcerated, mentally ill, or substance-abusing family member; domestic violence; or absence of a parent because of divorce or separation). ACEs have been linked to a range of adverse health outcomes in adulthood, including substance abuse, depression, cardiovascular disease, diabetes, cancer, and premature mortality."[64]

Ellen Goldstein, MFT, PhD, reported on the extremely significant increase in health risks for test takers answering "yes" to four or more of the ten questions on the ACEs questionnaire, saying "that 37 percent of those in an ACEs study screened positive for PTSD, 42 percent reported four or more ACEs, and 26 percent had elevated scores on both measures. Primary Care-PTSD and ACE scores were strongly positively correlated."[65]

The questionnaire and the analysis can be found on the CDC's website, at www.cdc.gov/violenceprevention/aces/index.html.

The term *ACEs* originates from a study of the same name, based on seventeen thousand participants and published by the CDC and Kaiser Permanente. The ACE study considers ten categories of childhood adversity. The study found: ACEs are common across all socioeconomic and culture/ethnicity lines; the accumulation of multiple ACE categories has a powerful impact on public health; and adverse childhood experiences causes trauma that tends to be held in the body, leading to mental, physical, and behavioral health problems throughout the life course.[66]

According to the Office of the California Surgeon General, "A robust body of literature demonstrates that ACEs are highly prevalent, strongly associated with poor childhood and adult health, mental health, behavioral and social outcomes, and demonstrate a pattern of high rates of intergenerational transmission."[67]

California's surgeon general, pediatrician Nadine Burke Harris, states, "Childhood trauma isn't something you just get over as you grow up . . . The repeated stress of abuse, neglect and parents struggling with mental health or substance abuse issues has real, tangible effects. This unfolds across a lifetime, to the point where those who've experienced high levels of trauma are at triple the risk for heart disease and lung cancer. An impassioned plea for pediatric medicine to confront the prevention and treatment of trauma, head-on."[68]

Now, we have a better understanding of what directly causes this physical disease. "Early childhood trauma resets your nervous system, your limbic, or your emotional brain; the amygdala is part of that," agrees Dr. Amen, who has worked with thousands of trauma patients. "Children who grow up in a stressful environment, it changes their microbiome in their gut. And when your microbiome is not healthy, you have more inflammation in your body and we know inflammation is one of the major causes of depression, of arthritis, of autoimmune disorders, and dementia. Stress causes many diseases. It also causes people to have trouble with their appetite and so they often overeat. You know, we have a country where seventy-two percent of us are overweight. Forty-two percent of us are obese—early childhood trauma increases that risk. And now we know that being overweight increases inflammation and stores toxins in your body, disrupting hormones. [Trauma] is just a risk factor for every bad thing."

"There was a longitudinal study that was done over time," psychotherapist Robbyn Peters Bennett told Jamie in an interview. "One of the largest ever done on how early stress in childhood affects physical health into adulthood. It was linking dysfunctional parenting and stress of a child to things like greater likelihood to die ten years earlier or a greater chance of being addicted to IV drugs. In fact, there were, like, ten ACEs and if you got six or more, that pretty much correlated to IV drug use, but if you had zero ACEs, you were essentially not going to be addicted to drugs. So, you begin to realize there is this correlation that was predictable and was so devastating. What they found is that ACEs are the number one reason for poor quality of life and death in the United States. This is shocking. Shocking."

To make the point of the damage caused by an overactive sympathetic, ACEs is brought up here because the allostatic load that builds in a child of deep poverty or long-term abuse will likely develop into PTSI, and then other physical illnesses follow.

A 2011 report by Gary W. Evans, Jeanne Brooks-Gunn, and Pamela Kato Klebanov outlines "evidence for a new, complementary pathway that links early childhood poverty to high levels of exposure to multiple risks, which in turn elevates chronic toxic stress. This cascade can begin very early in life. Even young babies growing up in low-income neighborhoods already evidence elevated chronic stress."[69] *Neuropsychopharmacology* reported in 2015 that "post-traumatic stress disorder (PTSD) is considered a disorder of recovery where individuals fail to learn and retain extinction of the traumatic fear response. In maltreated youth, PTSD is common, chronic, and associated with comorbidity."[70]

Jamie, who hit multiple aspects of the ACEs profile before turning eighteen, says, "I have so many stories of PTSI as an adult, but this one symptom is telling . . . I had a massive business failure almost twenty years ago that changed the trajectory of my life. I've realized now that the failure was caused by my sympathetic nervous system being messed up, my constant focus on the sense of doom. This false sense of doom that my sympathetic was creating, I think it caused me to *create* doom. If you're constantly stuck in doom mode, then doom is going to happen, whether it's in business or your family life or your relationships. I believe this is true for all of us. This reset has given me the possibility to take responsibility for and understand confusions in my life."

Nadine Burke Harris has done much to further the research and subsequent interventions around ACEs. In her book *The Deepest Well: Healing the Long-Term Effects of Childhood Trauma and Adversity* (Mariner Books, 2019), she discusses not only how ACEs are universal, across incomes and groups, but also generational, perpetuating the cycle of adversity. Further, she uses case studies and neuroscience to illustrate how what she calls *toxic stress* becomes rooted in our DNA.

So, does this mean we can pass our PTSI and the subsequent physical diseases to our children? "The generational issue is fascinating," says Eugene. "If you score high on the ACEs test, for instance growing up in squalor and dealing with violence, that kind of trauma can change your biology. The biological change overactivates the fight-or-flight mechanism. If the trauma is extreme enough, it can affect the DNA. This means, for example, people who survived living in concentration camps are likely to have children with a higher risk of health issues related to PTSI, like early heart attacks."

The idea of multigenerational trauma, or epigenetics, is not new. There are multiple, credible studies and books published on this topic, including *The Developing Genome* by David Moore in 2015.[71] The CDC has studied these phenomena for decades, stating, "Epigenetics is the study of how your behaviors and environment can cause changes that affect the way your genes work. Unlike genetic changes, epigenetic changes are reversible and do not change your DNA sequence, but they can change how your body reads a DNA sequence."[72]

Eugene says, "I think the whole concept of multigenerational trauma is just so crazy but well documented. The ramifications are huge. If you don't stop the trauma and deal with the issues already in place, then you'll keep passing it on.

"The DNA being . . . transformed . . . by big f-ing trauma, like three generations of families who go through terrible events, will birth a fourth generation prone to physical illness and the symptoms of post-traumatic stress. In my own family, my grandparents and their parents suffered two pogroms, my great-grandparents were beaten and killed. My father dropped bombs on Germans in World War II before I was born. Generations of trauma that led to an effect on the DNA, passed down. Epigenetics is now accepted in scientific circles. And three generations of big trauma means a genetic disposition toward poor health and developing PTSI becomes easier with this history."

Eugene believes that the PTSI symptoms he personally experienced through adulthood, culminating in the fugue states that he described to Dr. Porges, were in large part due to the multigenerational trauma his family endured. "Fascinating, right? Being born after generations of trauma does not *guarantee* you will develop a condition, but it does increase the likelihood. And I did *not* have a nurturing childhood that could have offset this trauma in my DNA."

COST OF TREATMENT IS A SOCIETAL BURDEN

The definition of PTSI has evolved, making it difficult to pinpoint a precise estimate of prevalence, yet the impacts on the national healthcare system and the criminal system are clear and devastating.

According to the *European Journal of Health Ethics*, "Individuals with PTSD seem to suffer from far more impairments in their general health conditions and incur many more costs than average insurants . . . In the control group, costs for mental disorders account for 19% of total costs. Costs increase by 142% in the year after an incident diagnosis of PTSD but return to the initial level 2 years later. Still, costs are at least twice as high in every year as in those for the comparison group."[73]

In the States, up to 30 percent of veterans return home from deployment with PTSD, according to the US Department of Veteran Affairs. Rates increase with extended and repeat deployments, among other factors. Over one million veterans currently receive disability benefits for PTSD, the third most common reason for disability claims (following tinnitus and hearing loss).

The VA also reports that at least six out of one hundred citizens (nonmilitary) will have PTSD in their lifetime, with at least fifteen million adults suffering from PTSD any given year. Because living with PTSI affects so many aspects of daily life, the impact on quality of life is tremendous.

The Veterans Disability Benefits Commission in 2007 reported that veterans have the poorest overall healthcare and quality of life, with one in three veterans unable to work at all, and those statistics have only increased.[74] Unfortunately, many veterans wait years to seek treatment. The cost of treatment is a burden, both to the patient and the healthcare system. At the same time, lack of treatment carries an unfathomable price tag. Not counting the social and emotional costs, a Rand study found that losing soldiers to PTSI and suicide costs between $4 billion and $6 billion over a two-year period—and that was back in 2008.[75]

The prisons are full of inmates with PTSI, as discussed earlier. In combination with traumatic brain disorder, which often removes the brain's ability to access logic, the impulsivity of those who suffer from PTSI has created the perfect storm, making the decision to be involved with something illegal, even dangerous, very easy. In a 2017 CNN special investigation, *War Crimes* by Kaj Larsen, Colorado Springs was discovered to be "ground zero in what may be an approaching tsunami: the alarming rise in soldiers with PTSI arrested for violent crimes. Traumatized by war, a growing number of these vets find themselves unable to leave the violence on the battlefield. The skyrocketing suicide statistics and increasing crime rates among veterans are just sentinel warnings of PTSI in the masses."

Those with PTSI must be offered relief before the impact of crime and suicides devastates more families and the healthcare system collapses.

"I want to make it clear: Dr. Lipov and his DSR are a possible solution to these problems," says Jamie. "I do not say it lightly when I say that this is the greatest human innovation since the advent of penicillin. You get it, you may be saved."

CHAPTER EIGHT

Reversing the Injury

In 2016, Matt Farwell wrote an article for *Playboy* magazine about his experience with DSR. "I would be pissed I didn't get this shot earlier—if I weren't so grateful I got it at all," he wrote.

In this chapter, we will look at what DSR can do for Matt or anyone suffering from PTSI symptoms, especially when the procedure is done in conjunction with other medicines and modalities. We will also appraise the value of popular modalities that have become synonymous with the fight against post-traumatic stress, addressing how many of these are useful but only truly effective after the injury has been corrected, allowing the emotional healing to begin.

Matt had come back from sixteen months overseas as a combat infantryman only to be diagnosed with severe PTSD. He spent seven years trying to cure his insomnia, violent outbursts, and suicidal ideation via multiple modalities with no luck. Until he came to the Stella Center.

"It turns out a big part of the cure was under my nose the whole time," he wrote. "Well, six or seven inches under my nose and a couple of inches back and to the right, in a cluster of nerves by the spinal column called the *stellate ganglion*. Two injections of a couple of local anesthetics . . . and I was pretty much back to myself. Dr. Eugene Lipov . . . tells me the Navy SEALs call it the God Shot."[76]

Eugene's DSR has finally gained traction. The quick, available procedure for long-term relief with minimal side effects, as well as high clinical efficacy rates, has made DSR an attractive approach for both physicians and patients. PTSI and the majority of ailments or issues linked to the injured sympathetic nervous system discussed in chapter seven no longer need to be a source of crisis for the nation or for individuals.

"[The] utility of stellate ganglion block (SGB) seems to be expanding rapidly at this time," states Ken Candido in an article written with Eugene, referencing a report by Dr. Hong-Ying Zhao, et al, who recently demonstrated a marked impact of stellate ganglion block on chronic ulcerative colitis (an autoimmune disease) on his patients, relieving abdominal inflammation and pain.[77]

One common argument against the treatment is that it is not yet FDA approved—but that argument is a moot point. The Department of Veterans Affairs Health Services Research and Development Service published an article in 2017, stating, "Ropivacaine and bupivacaine are FDA-approved for production of local or regional anesthesia for surgery and acute pain management, including in the head and neck area. Injection of these drugs into the stellate ganglion for PTSD is considered an 'off-label' use for a different disease than described in the drug label, which is legal."[78] This isn't a rare or odd occurrence; everyday staples like vitamins and baby formula are also not FDA approved.

In a December 2021 abstract, Eugene and two other doctors wrote, "Traditionally, SGB has been used for sympathetically mediated pain conditions such as complex regional pain syndrome, shingles, and others. Recently, SGB has been described to be effective for other indications such as post-traumatic stress disorder symptoms and menopause-related hot flashes, as well as ulcerative colitis with accompanying reduction of interleukin-6 . . . [as well as] an unanticipated finding of PASC symptom resolution."[79]

THE BEST OUTCOMES

The best outcomes only happen when, as Jamie likes to say, "The hardware is fixed and then the software." Using other modalities or treatments without getting DSR is simply fixing the software. Or, to mix metaphors, applying a Band-Aid. He says, "These modalities can be fantastic but typically more effective after DSR treatments. Then, the patient will have clearance to do the emotional work. Otherwise, it is outside-in healthcare. You need to deal with the root cause first."

Yet, even when DSR successfully reverses PTSI symptoms, that does not mean the work is done for the patient. The memories, emotions, ingrained responses, and habits . . . they are all still there. In 2021, Eugene worked with a patient who had been forced into the sex trade as a child; her PTSI symptoms eventually caused severe health issues, including painful rashes, adrenal failure, and insulin resistance. "My body was falling apart along with my mind," she says. "DSR gave me my body back. But it didn't make my past go away. I still have feelings, emotions, memories, belief systems, and broken relationships to work through. The injections just ensured me that I can now do that. And it makes it a hell of a lot easier, not having to deal with physical pain and sickness on top of the emotional stuff."

This patient went on to find further success with additional approaches. "After doing the procedure with Eugene, the intrusive critical thoughts completely disappeared and I could get a lot farther in therapy than ever before. I can see different therapeutic skills actually work now. I have been able to add every single food back into my diet, with the exception of tomatoes. I am able to calm down quickly whenever I get triggered. That was literally not an option for me before. I had no capability of homeostasis and now when I see/feel it happen I get a rush of joy."

The most common modalities, practices, or medicines deal with specific symptoms, often successfully, yet can mask the core problem. According to Eugene, "In my experience, eye movement desensitization or mind-body practices such as yoga and meditation can be helpful for some PTSI patients. And for some patients, pharmaceuticals *are* effective, even if I don't like most pharmaceuticals. If the medical evidence is there, that it's working, then use it."

Patients can work with their doctor to develop a treatment plan that combines conventional, targeted treatments with alternatives; the key is customizing the treatment plan to the patient.

THE "GOLD STANDARD OF CARE"

In-depth clinical research on post-traumatic stress didn't really go big until the 1980s, and most of that research continues to be unfortunately focused on the military. Often, even the best doctors still struggle to help those suffering from post-traumatic stress, unaware of Eugene's dual sympathetic reset procedure.

Instead, they are trained in today's "gold standard of care"—pharmaceutical treatment plus cognitive therapy—which alleviates the symptoms in some patients but not all and often requires long-term use of prescription drugs.

The bulk of current treatment guidelines in mainstream medicine can be classified as pharmacological therapies or psychotherapies.

Prescribed medications generally fall into the categories of antidepressants, anti-anxiety medications, and prazosin (for suppressing nightmares), none of which should be taken without consulting a doctor first, and then need to be taken consistently, following the doctor's orders. Many of these can help manage specific symptoms, such as anxiety or insomnia; it should be noted, however, the pharmaceuticals involved are selective serotonin reuptake inhibitors (SSRIs) that may reduce anxiety or arousal but don't address all of the symptoms. Or they are antipsychotics, which may come with very serious side effects.

Psychotherapies are geared around developing stress management skills and behaviors, and often include cognitive behavioral therapy (CBT) and cognitive processing therapy (CPT), which are based on talk therapy and recognizing cognitive patterns that are negative or repetitive. Prolonged exposure therapy has the patient closely assess traumatic memories or nightmares in order to learn how to process them calmly when they arise, with the final aim that exposure will take away the fear. Hypnotherapy requires a licensed practitioner to help patients walk through closed-off memories. There is also eye movement desensitization and reprocessing (EMDR), which combines exposure therapy with guided eye movements that are meant to change the way the patient processes the trauma. While these approaches have become fairly mainstream, it is important to do research on any treatment you choose and to consult your doctor.

The US Department of Veterans Affairs has finally mandated that all veterans treated for PTSD (the government has not yet changed the designation from a disorder to an injury) have access to either prolonged exposure therapy or cognitive processing therapy, not just prescription drugs.

"However, veterans remain reluctant to seek care," states Eugene in a recent article, "with half of those in need not utilizing mental health services. The long duration of time and consistent follow-up required for these treatments, coupled with the social stigma of seeking psychiatric care, often hinder both civilian and military patients' desire to seek and continue treatment."[80]

Eugene then goes on to reference statistics from Dr. Charles Hoge, stating how among veterans who begin PTSI treatment with psychotherapy or medication, dropout rates are reported to be as high as 20 to 40 percent in randomized clinical trials and considerably higher in routine practice. While therapeutic efficacy is reported to be achieved in 60 to 80 percent of patients compliant with medical treatment, intention-to-treat analysis shows efficacy decreases to approximately 40 percent when accounting for patients not completing treatment.[81]

ALTERNATIVE APPROACHES FOR TREATING PTSI

DSR is not the endgame—getting healthy is the endgame.

Many alternative approaches have seen varying levels of success. The key is to make sure the patient and doctor are doing their due diligence with research and pay attention to outcomes.

"If a treatment that is not shown to be efficacious is nevertheless delivered to veterans, and if the treatment is relatively inert, even if it does not harm the veterans, it may demoralize the veteran," said Richard McNally, a Harvard University psychologist and post-traumatic stress expert. "Providing treatments that do not have a good basis in evidence can result in people not improving, therefore getting demoralized and therefore not seeking treatment that can actually help them."[82]

Just because a modality or medication has become popular, that does not mean it is truly effective. On the other hand, there *are* many promising approaches currently being researched for PTSI patient application, either as stand-alone treatments or as supplementary care. Again, the key is to work with your doctor to find safe and effective approaches that work for you.

Ketamine treatments, which is an anesthesia drug, are one such practice that has found acceptance within the medical establishment. However, for ketamine to work, it typically takes four to six infusions to have any real effect on PTSI, and the efficacy may be short. It has been shown to offer improvement in a number of other medical conditions, such as depression and anxiety, though it may be beneficial for patients who have clinically diagnosed PTSI and are always in the fight-or-flight mode to start with DSR. Eugene states, "We have a close working relationship with a ketamine clinic and often have meetings with physicians there to decide which technique is best for a particular patient, or if perhaps both are necessary."

Psychotropics, THC, and psilocybin are becoming more popular as treatments for PTSI but are still being studied. Psilocybin has been recently found to be effective in cases of clinical depression, but no major studies have yet been published in the area of trauma relief, and though it can be easy to access, it is extremely important to talk to a medical professional before jumping into this approach. The same can be said of the recent attempts to use the street drug ecstasy to treat PTSI symptoms. The FDA is close to approving a clinical protocol that combines ecstasy, technically MDMA (methylenedioxymethamphetamine), with talk therapy sessions. Regarding THC, as noted earlier, Amen and Faber have found that consistent THC use ravages the brain over time in their work with some patients.

Ayahuasca, a psychoactive beverage made from plants found in South America, has recently gained popularity in the US, though it is still largely illegal. Some partakers say they have experienced relief from their symptoms caused by post-traumatic stress. Evalee Gertz, one of Eugene's PTSI patients with a childhood history of trauma, is leery of new ayahuasca practitioners with little to no medical training or practical experience with the potent drug, though she has traveled to Peru multiple times to work with authentic indigenous shamans who have access to centuries of tribal knowledge around the healing properties of the ayahuasca plant and focus on releasing both physical and spiritual toxins.

In a 2022 interview with Jamie, Evalee describes her experiences at the ayahuasca retreats as "extremely beneficial," opening her up to life-changing epiphanies about herself and the world, referring to ayahuasca as "a powerful tool. Though not a solution in and of itself, it can really help you to see how you're holding yourself back or see what your blocks are." She defines the outcomes as highly individualized, personal, and "very intense, so it's not something to do lightly, you need to make sure you're in a safe environment with people who know what they're doing and can take care of you."

At a retreat in early 2022, Evalee came to the realization that the constant state of stress she managed originated not from her consciousness but from a place she couldn't control. Around that time, she learned more about the symptoms arising from post-traumatic stress injury, realized she had PTSI, and sought out Eugene. After undergoing DSR at the Chicago Stella Center, she found that her anxiety was suddenly manageable, as were her emotional reactions. She was calmer, and the modalities or practices she had incorporated into her life were now even more impactful.

Evalee has been working on herself for years, utilizing various avenues. "Honestly, I think meditation is one of the most powerful practices," says Evalee. "It's part of a daily routine that keeps me sane and [I believe] that everyone should do it. Meditation helps you to be present in your body. They have done studies to show it literally rewires your brain. It helps with anxiety, helps raise your IQ level . . . It's such an important human modality, sitting and connecting with your breath and being in your body and letting go of the external world, and going inward and giving yourself permission to just be and not do and give yourself permission to just look at yourself instead of trying to fix or heal. I'm a doer, so sitting and just being can be difficult but really a very healing thing. I started meditating when I was nineteen . . . It helps with stress. It helps clear the brain, allows that constant chatter to turn off."

Traditional mind-body care practices are now being researched more closely for their application in reducing trauma-related symptoms. Yoga and meditation and bio-energy techniques such as qi gong, Reiki, distant healing, and acupuncture are generally viewed as safe, having been practiced for centuries.

Other traditional healthy mind-body approaches include neurofeedback, music, animal-facilitated therapy, art, dance/movement, massage therapy, exposure-virtual reality, and spiritual ministry. None of these seem to be harmful and often lead to stronger bodies. The only issue is that they focus on one or two symptoms and, when used singly, rarely seem to have high efficacy for long periods of time.

Brainspotting is a psychotherapy alternative practice linked to EMDR. The original EMDR treatment is only conditionally recommended by the APA, with an efficacy rate of approximately 25 to 33 percent. Brainspotting, a recent innovation, seems to be more effective, involving a body activation experience that happens when a patient reveals a traumatic event via a resonating spot in the visual field. This treatment allows the patient to choose what issue to focus on with minimal guidance from a counselor, who is paying attention to where the eyes focus; it's not as widely known but both utilize bilateral stimulation.[83] The research is not yet clear on how either specifically works.

"It just makes sense to bring all potential therapies to bear," stated the former head of the Defense Centers of Excellence for Psychological Health and Traumatic Brain Injury, Army Brig. General Loree Sutton.

The general was instrumental in bringing to life an alternative treatment known as the Theater of War, which focuses on a performance based on two plays from ancient Greece dealing with the ramifications of battle. The dialogue was translated straight from the original Greek, but the words and images evoked are believed to "speak" to combat veterans of today. In fact, writer Bryan Doerries was inspired to produce the performance after discovering that the Greeks used theater as a way to provide catharsis for combat veterans and reintroduce them into society, which is discussed in Dr. Jonathan Shay's book *Achilles in Vietnam*.

The approach has not been found to have produced significant results, but there should be more research in this area—as well as more out-of-the-box thinking when it comes to treatments.[84]

PTSI treatments have come a long way, but none have resulted in the same efficacy numbers as DSR.

TRAUMATIC BRAIN INJURY

We are bringing up traumatic brain injury again because so much of the care goes hand in hand with also treating post-traumatic stress injury.

A TBI, caused by a sudden jolt to the brain that disrupts function (categorized as mild, moderate, or severe), involves damage to the brain *structure*. On the other hand, the overactivation, or "injury," to the nervous system that results in PTSI affects a brain *system*. The structure and the system injuries do overlap, however; both are physical injuries and both are root causes for the symptoms that a host of medicines and modalities try to solve. And if a patient is suffering from both PTSI and TBI, they will need to address both issues in order to find full relief from the trauma symptoms.

However, the vast majority of people who have suffered a blow to the head, or brain tissue damage due to overuse of drugs or alcohol resulting in a TBI, did not have access to DSR, or even knowledge of DSR, within that time frame. And "growing evidence and acceptance have shown that PTSI can occur after TBI, but the reported prevalence rates vary widely by study from 0 to 56 percent."[85]

According to the American Speech-Language-Hearing Association, roughly 5.3 million Americans are suffering from a traumatic brain injury–related disability. Worse, the CDC has said 57,000 TBI-related deaths occur each year in the United States. Roughly 2,500 of those deaths are children.[86] "An estimated 75 percent of TBIs are considered mild or moderate; however, even these seemingly innocuous injuries have been linked to increased risk of Parkinson's disease and dementia," states Research!America. "Severe TBI can cause lasting damage, ranging from consistent headaches to chronic and debilitating cognitive or movement impairment."[87]

This is an important discussion because PTSI patients who also have a traumatic brain injury should be working on both.

At the Amen Clinics (now spread across the states), they work to uncover the root causes of the trauma or the injury. Dr. Amen says, "We believe it is critical to look at your brain within the context of your life, which includes biological, psychological, social, and spiritual influences." The clinicians use in-depth interviews and other assessments in order to understand the needs and issues of the individuals. Then, "unlike traditional psychiatry that rarely looks at the organ it treats," the clinics use brain SPECT imaging to measure blood flow and activity in the brain to more accurately diagnose the issues (concussion, alcohol or drug damage, ADD, depression, etc.) and create a combined holistic/pharmaceutical plan that is biologic-physiologic-social-spiritual based. There is additional brain mapping and neuropsychological testing to measure cognitive, emotional, and intellectual functioning, as well as diagnostic testing for contributing biological issues.

The goal is to create a treatment plan specific to the individual brain type, addressing root issues versus the symptoms.

Treatments include (but are not limited to) therapy, targeted integrative medicine, and nutritional supplements.

"The same treatment plan won't work for everybody and, in some cases, it could make your symptoms worse. The Amen Clinics database, with over 190,000 brain scans, helps us more successfully identify various types of . . . mental health conditions, so you get the answers you've been looking for," claims the clinic's website, at amenclinics.com. This is an incredible data set to work from, in terms of seeing what is corollary and what is causal and in terms of brain health and mood function.

There is a binary relationship between Eugene's work and Dr. Amen's and Dr. Faber's follow-up work around "brain health" using scans to determine the need for natural supplements, diet, and lifestyle changes in order to restore proper blood flow in the brain. The effects of trauma have long been misunderstood and stigmatized as mental problems. Developing brain health through Amen's holistic work along with resetting the sympathetic nervous system are the foundational and long-term restoration that the human race sorely needs.

Jamie states, "I don't think most patients are able to experience the benefits of the sympathetic reset until TBI is resolved if they have both. Severe TBI appears to blur the patient's ability to experience the benefits of DSR, even if a spouse or child sees massive changes in the person.

"My observed clinical data leads me to believe that the ten to fifteen percent of patients who don't exhibit overt benefits from the dual sympathetic reset—which can happen for multiple reasons—is often because of TBI, and that needs to be addressed. Dr. Amen could have a patient on supplements for six months and maybe do a series of hyperbaric chamber treatments, which can dramatically reduce TBI. But that patient can still have an overactive sympathetic nervous system, so the PTSI needs to be addressed. The patient goes in for DSR, the sympathetic nervous system is reset, and they are going to be ready to take on any other mental health or emotional battles they may still need to address."

He goes on to say, "With the massive leap in relief experienced by DSR patients in just a couple of days, it might be the boost patients need to help them through the medium-term holistic modalities that Amen proposes. I believe that Eugene has created a spotlight on the pioneering work that Amen has been doing for over thirty years, with such an immediate result it allows us to understand Amen's innovation, which takes longer to see."

DSR ISN'T FOR EVERYONE

The authors of this book are not touting the dual sympathetic reset as a universal panacea.

First of all, there are financial considerations for some—a harsh reality of the American healthcare system. Even if the DSR is ideal for a patient, health insurance companies don't yet cover it as a treatment for PTSI (though the SGB is covered for other medical conditions, like nerve pain). The treatment generally costs roughly $3,000. There are nonprofits that may be able to help with the fees (addressed in the final chapter), and it is also worth it to talk to the local VA if you are military. "It is my hope that this procedure will soon be available to patients of all economic means," states Eugene.

But there are also health-based reasons to consider before choosing the procedure. For instance, the procedure may not be recommended for someone suffering from heart or lung disease or who is taking blood thinners, which are contraindications. Using an ultrasound machine can be challenging on a patient who is morbidly obese, as it is hard to place the needle correctly while administering the DSR, so it is recommended that such patients who wish to get the injections embark on a weight-loss plan under their doctor's guidance to ensure their safety.

Further, there is as of yet no robust data on patients who are under twelve or ages eighty and up. It is not yet possible to make a well-informed decision about using the dual sympathetic reset procedure on these patients.

Finally, 85 percent of DSR patients report relief, but there are people who report no difference or very little impact on their symptoms. In the end, certain individuals have a history and biochemistry that may leave them impervious to the injections. If a patient does not find relief with DSR, Eugene encourages the patient to reengage with their doctor and formulate a new plan of attack. There are other options out there.

WE NEED THEM TO BE ALIVE

In 1925, back when antibiotics had yet to be invented (1928) and indoor plumbing was still just a dream for many Americans, the stellate ganglion block was first used on an asthmatic. The block worked; the asthma was improved. The immune system was reset. Yet, here we are a hundred years later and the medical establishment is only now starting to embrace SGB/DSR beyond that of a simple nerve block, despite so much evidence supporting the wide array of individual and social ailments that can be helped by this procedure.

From misery, to illness, to suicide, PTSI no longer needs to drive patients over the brink.

"We need somebody to be *alive* in order to be able to help him," Dr. Carrie Elks at Fort Bragg told Jamie in 2021; Elks is a specialist in military mental health specifically treating PTSI or psychological trauma for the past ten to twelve years.

With the Stella Center opening clinics in cities across the US and the world, access to relief from trauma is becoming a reality. "In the really extreme cases, when somebody is homicidal or suicidal, there needs to be an immediate resolution," states Elks. "It's good to know there's something [like DSR] that we can do to get them to a baseline, so that they can get the treatment they need."

For Corey Drayton, the filmmaker who endured a lifetime of micro and macro trauma that led to symptoms including suicidal ideation and eventually colorectal and prostate cancers, the DSR injections were nothing short of a miracle, saving his life on many levels. The procedure "lowered my defenses, my resistance to the things that were actually necessary for me to do my own healing. I moved through waves of intense emotion over the course of the recovery period the first two days, including grief, sadness, rage . . . but eventually arrived at a place of inner quiet and focus. Over the following month, I did feel the weight of a lifetime of emotions that I hadn't yet felt or processed related to my life traumas. It did eventually become so heavy that, for about three weeks, it was all I could do to haul myself out of bed. I felt overwhelmed by my internal reality. I couldn't take calls. I couldn't work. I couldn't visit with friends. I couldn't leave the house. Uttering the simplest of words felt like some Sisyphean task.

"That's when it really dawned on me what DSR is: It's the catalyst. It starts a chain reaction that improves the conditions for healing through other modalities. I found then that I was completely present for the deep therapy I had done so often in the past that never seemed to penetrate into that deep wound I was carrying. I have found that I am finally feeling the full depth of my emotions, I am not in the place of numb, subdued disconnection."

Corey believed the dual sympathetic reset was "perhaps the fundamental tool necessary to create the conditions for healing, but it doesn't absolve you of responsibility. It's not as simple as getting DSR and then going back to the same bullshit routines and patterns that lead to your crisis point. You must get yourself to a place where you operate differently after having DSR—with greater intentionality, with greater presence, with greater honesty. This must become cellular, like an instinct. Without this work, I fear one wouldn't derive the full benefit of having a DSR."

Those benefits are worth the effort for everyone, from those suffering from the darkest manifestation of trauma symptoms to those living with manageable—but unnecessary—unhappiness or malaise.

Evalee (Eugene's patient mentioned earlier in the chapter) had been cynical about the veracity of DSR but decided that since it was safe, it was worth a try, especially if it could help her address symptoms that she seemed to have no control over. "I didn't have any idea of what to expect because people who have really severe PTSD, like they've come back from war or whatever, and they're literally disabled, having a hard time functioning because of it, that is super not me—I am a very high-functioning person. So, it was like, I don't know how I'm gonna feel or what's going to be different. I was hopeful but also kind of skeptical."

But once she and her fiancé underwent the procedure and a few days passed, it was evident they were processing their interactions with each other and the world around them with more clarity and less reactivity. Evalee stated, "I am grateful for a tool like DSR, which makes dealing with life easier. I noticed that now, when I'm feeling reactive, and I'm having, you know, those feelings come up and those anxious thoughts in my head, [it is] so much easier to kind of talk myself off that cliff."

On a bigger scale, she was also grateful that DSR exists as a tool for others and agrees with Corey that the simple biological procedure is not the end of the healing journey. "I believe everybody has some basic level of trauma. I believe that the issues of the world could be solved if everyone was committed to doing work on themselves. Honestly, I think that the whole point of being alive is to grow and evolve . . . but doing that kind of work is really hard and really scary and can be really painful and most people will do anything to avoid it. You know, everyone's always pointing the finger outward and trying to change the circumstances around them in order to feel safe and okay. You've got to point inward. DSR gives you the chance to do that."

These are the types of reactions that led Eugene Lipov to push for more acceptance across the medical community. It is also why Jamie claims that, as knowledge of the procedure enters into the public zeitgeist, the dual sympathetic reset could change how humans move through the world.

CHAPTER NINE

Everyday People

How much trauma can the human mind sustain? That answer is different for everyone, dependent on so many elements, including the type and length of the trauma, the brain health going into the trauma, genetics, and availability of medical and emotional support systems.

"I think the real question is, how do we decrease that trauma in the most efficacious way possible?" states Dr. Faber. "Some trauma is not necessarily bad—it can challenge us in good ways sometimes—but how do we create a better trauma-managed society? Because trauma is not gonna go away. How we manage it becomes extremely important. We have a long way to go on knowing exactly what to do."

Unfortunately, we can't manage our trauma fallout until we recognize it in ourselves.

In this chapter, we'll look at examples of several real, everyday people who have experienced trauma. Who are struggling with what they consider everyday, unavoidable health or mental issues. People you work with or who live on your street. People who perhaps resemble a family member. Or you. People who could change their lives with a fifteen-minute procedure.

A FIREFIGHTER

Volunteer firefighter Rose knew about post-traumatic stress intimately before ever donning the uniform.[88] Her rural fire district, on the other hand, stands by a culture found in firehouses and police stations across the country, a culture based on "I'm fine, I'm not weak," and generally turning a blind eye to those on their front line, even those likely affected by the repetitive trauma they encounter. Rose was surprised at the lack of resources made available, especially to the volunteers. "Resources *do* exist, they are just not well known or readily accessible—even worse, not comfortably accessed by someone who is deep-rooted in the stigma around how firefighters are tough so 'we should be able to handle this.'"

She knew the signs of PTSI in the firefighters around her, and eventually within herself, having raised a child with post-traumatic stress. "We were in and out of doctor's offices, watching her from the age of nine go through all of these stages of PTSD, which morphed into self-destruction, any coping mechanism she could come up with. I paid the most expensive child psychologist in the city and commuted out there every week for four years. And we never got to the bottom of it."

Rose and her daughter are both now looking into getting the Lipov procedure. "I'm more intrigued for my daughter than even for myself. To see my daughter finally find peace . . . And now, because I was a bystander to her and all she has gone through, I do recognize exactly what I am going through now, and the firefighters around me, whether they're acknowledging it or not."

With the understanding that an event may be traumatic for some but not for others, she believes many of the first responders who work with her have trained themselves to abide by the cultural norms and not to "see" the post-traumatic trauma and the effects. They may even be fine at the time of the event, but then flashbacks and other symptoms crop up weeks or months after the trauma. Perhaps someone becomes a new father, and the death of an infant he worked on a year ago is now applicable to his life, and he begins to have panic attacks. Or the shock of the event wears off and that which was dormant comes to the forefront.

For example, a week or so after witnessing the aftermath of the suicide of a teenage girl who was the same age and build as her daughter, Rose came home, entered her garage, and found family members engaged in a project that had spattered blue paint on the walls and left blue footprints on the floor. It looked like blood spatter. It looked like the scene of the suicide with the dead teen, right down to the footprints. "It wasn't blood, but it might as well have been. I'd kept it together up until then, but I lost my shit that night."

The suicide had been the fifth fatality Rose was called on scene for that week. "There was an eleven-year-old girl, another a twenty-year-old kid. There was a healthy thirty-three-year-old I worked on all the way to the hospital. Everybody was far under the age of seventy. I didn't know what to think. I'd be driving, not knowing where I was going, unable to focus. I couldn't process anymore. But no one offered me any kind of debrief or help. The only post-incident conversation I had was with my partner, who'd been on the call with me. The lack of follow-up was concerning."

There was a poster on the wall of the firehouse that Rose thought might lead to a hotline or tele-health counselor, since it offered a 1-800 number under the image of a distraught ambulance driver—but it was for a drug and alcohol abuse program.

Eventually, she found a personal trauma specialist outside of the district. There is a critical incident response unit that can be accessed by the county, and there are critical incident classes available by the state and county, but none of those are offered or advertised by her firehouse or any of the houses in the county. She is currently working with administration to create new policies to correct that, as well as building a curriculum to teach trauma-relief classes to her peers. "No one is getting the help they need. Resources simply don't exist for the first responders. I don't wanna bash anybody with that because I think everybody does the best they can with their limited knowledge of PTSD, but that definitely needs to change. Now."

Humans can have this tripped in us by being abandoned by a father or mother, or just the weight of emotional stress from being around unhappy parents. The invisible machine does not think, it only interprets and acts. What specifically trips it is a little different for everyone because people and their DNA, general health, emotional support systems, and experiences are different.

A SOCIAL WORKER

Leanne will tell you that she's fine—most of the time. She believes her issues with constant anxiety and sense of foreboding stem from growing up in a house ruled by an overly critical mother but did not see her struggles as anything out of the ordinary. She was skeptical when her husband (who worked with Jamie on an earlier project) suggested she meet with Eugene and get the procedure, not believing she needed help with her nervous system or brain health.

"I feel functional but not optimal, I guess you could say," she states in an interview before DSR. "I felt pretty comfortable, physically, as a child. But I did feel things pretty intensely. I was incredibly shy. I didn't talk very much. I was constantly being told, "Speak up, speak up." Maybe a little bit fearful . . . but I don't think it was maybe abnormal for a child.

"Now, I'm pretty healthy, I guess, but I can easily go to a place of unhealth, depending on the circumstances. I'm not a hundred percent convinced I need [dual sympathetic reset]. I don't feel like I'm somebody who has had trauma in my life."

In the interview, she hesitates, looks at the floor, and then says quietly, "At the same time, I want to feel better. I feel like I could process the stressors in my life a lot better. I've done a lot of work mentally, and I'm in a healthier space than I have been . . . but at the same time, I don't feel like my body has gotten the memo. You know, when I was younger, it would take a really stressful situation lasting a really long time before I would start to feel physically ill, but I'm at a place now where I make that jump pretty fast. Something stressful happens, and soon after, I start to feel sick to my stomach. That pathway is so well worn. I guess I do worry about my long-term health. If this is an opportunity to restart that, then I welcome it."

Leanne is the perfect example of how so many of us experience persistent micro-aggressions for a long period of time and yet are unaware when the allostatic load has become too heavy and has tipped the sympathetic nervous system into a state of constant overdrive. We know it is important in our society to be perceived as mentally strong and competent, and so that is what we project, pretending to ourselves that the symptoms of PTSI are simply normal issues that everybody deals with.

But, then, Leanne underwent Lipov's procedure and everything changed. In a follow-up with Jamie, she describes herself at peace, no longer anxious, with no more stomach issues, saying, "It took fully six months for me to be able to look back and see the real impact. Up to that point, I didn't trust that the change was here to stay. But, now, I think it is. The biggest change is that my normal stress is just that, normal, rather than the extremes I used to feel. I don't go into panic mode anymore. I can handle [stress] like a normal person."

A PIANIST

While growing up in Ireland, Frank was sexually abused by one of the priests who ran the school system. Frank's younger brother and eight of his childhood friends died by suicide after also being victimized by this priest.

Now in his late sixties and one of Eugene's patients, Frank views himself as lucky. He moved to the US, married, and built a successful career as a piano player. However, he and his wife have attended counseling together for twenty years due to his struggle with depression and suicidal thoughts. He claims to have "come alive" on stage as a dueling piano player, but when he was home he spent his time alone, often drinking alcohol. Prescription Prozac did not help, simply leaving him numb.

Eugene examined Frank, reviewed his medical records, and administered the PTSD checklist. Frank was a solid candidate for the dual sympathetic reset. The pianist experienced almost immediate relief after the injections. "I felt more myself," he recalls, struggling to put his feelings into words. "I used to pretend I was okay. I didn't have to pretend anymore."

When he and his wife went to see their therapist, the therapist recognized the change in him immediately, asking, "What is going on with you? You're completely at ease! I've never seen you like this, Frank."

Eugene and his DSR procedure allowed Frank to achieve a baseline of mental health; his sympathetic nervous system was reset. Medication and talk therapy had been unsuccessful before the reset, but having his amygdala back at a pre-trauma state now meant he could find emotional healing through other modalities. He began studying meditation and performing Celtic music, and helping others learn mindful awareness.

A NURSE

"I've been struggling tremendously with my mental health for the past few years. Triggered by multiple losses and seemingly insurmountable grief, I'm a hope junkie who can't find a fix. I cannot remember my last tear-free day. I can't remember a day that wasn't in some way crippled by anxiety and fear. Med after med and so far nothing has worked. I'm stubbornly optimistic but I'm so tired and so scared I won't find my way back to myself. Sharing because I know I'm not alone in this but my brain plays tricks on me and I *feel* so alone in this. Always looking for a sign that everything's gonna be okay, when nothing feels like it will ever be okay again. But the signs are always there, so I'll just keep seeing them until I believe them again. Because someday I'll believe them again, right?"

Sara posted these words on Facebook in the winter of 2021. She is a nurse at a federally funded primary care clinic in Oregon, where they serve everyone regardless of ability to pay. The patient population leans heavily toward those who are on the fringes in the best of times. She and many of her colleagues were already struggling with how to cope with those with post-traumatic stress, and their own growing allostatic load, before COVID hit. The administration, part of a much larger corporation based in another state, did not acknowledge the trauma their staff was facing, except for a handful of motivational emails that started showing up throughout the pandemic.[89]

In the early days of chaos, her clinic manager and nursing supervisor were fired due to incompetence. Almost two years later, these jobs have yet to be filled. In response, half the staff quit. In February 2022, she said, "We went from having two medical assistants per provider with some to spare, to just one per provider. We had four nurses. Now it's just me."

Shorthanded, the other stressors became untenable. Like everywhere else, there was a major shortage in PPE; the medical staff were expected to reuse gowns and masks. "I was never personally afraid [of the virus]—I'm in the field for a reason and risk is inherent. But patients were freaking out. Lots of panicky phone calls, tears, anger. And protocol changed weekly. For the longest while, we had to have someone sitting in the hall all day taking temperatures and screening. Which is a farce and took away an MA or nurse from our regular duties.

"We had temporary supervisors coming in occasionally from other places in the organization. We had no trust in them; they had no understanding of our clinic or the needs of our community. They'd swoop in, slap our wrists, impose new dictates, then leave."

There was also a huge local uptick in substance abuse and suicidality, and Sara's clinic was without a mental health person, so that care fell to the nursing staff as well. "I don't mind; I have an MSW and that was my first career, but there's only so much time in the day."

Daily life suddenly meant enduring constant overwhelm, always feeling behind, never catching up. There was no support for her or her staff, much less positive recognition. Despite Herculean efforts, patient care suffered, as they were limited to mostly videoconferencing with the sick. Everything but COVID responses was neglected. There were horrifying gaps in treatment as all but the most serious of medical issues became "elective."

Sara knows the post-traumatic stress symptoms within her have reached a point that she has changed, and she needs help to get back to a good place. "Now? I'm scared all the time. This has definitely unleashed the effects of past traumas. I've always been go, go, go, succeed, succeed, succeed. I hate feeling incapable, like my best is not good enough.

"Instead of feeling nurtured over the past couple of years, I've felt scrutinized with a punitive tone. I mean not just me, all of us. Yet, though losing so much staff has created an unmanageable workload, the people who stayed are awesome and resilient, people with a good sense of humor and who really give a shit."

Sara does feel like her workplace situation is slowly getting better. Vaccinations have eased some of the burden, with patients showing their gratitude by offering gifts and tears, and there is kindness and hope alongside the negativity.

"Our clinic is starting to rebuild itself with compassionate, caring, and committed people. I just got to work to find a box of chocolates from an anonymous source. And my new nursing supervisor is an angel. So kind and supportive, always cheerleading and offering affirmation. I'm feeling hopeful that we will become stronger because of the last two years, a place where we look forward to getting up in the morning and truly become a family."

But the post-traumatic stress is not easing, at least not for Sara. "I've never experienced such brain fog. Even right now, I'm grasping for vocabulary, and words have always been my thing. I can't read. Can't focus. Can't watch anything but trashy reality TV or serial killer documentaries." She laughs but becomes serious again. "I was waking up every single morning in full-on panic attack as soon as I opened my eyes. My short-term memory is shot to shit still. Absolute exhaustion. Depression, helplessness. Can't even tell you how many mental health meds I've tried. I stopped drinking one month into the pandemic, which wasn't maybe the best timing, but I knew I'd end up relying on it too heavily.

"I've been coming home with less than zero left for my family or myself. At first, I tried to keep up with acupuncture, Reiki, therapy . . . but that took too much energy. So, I've been white-knuckling it."

Luckily for her community, Sara is a bright and engaged caregiver, working for the betterment of herself and her patients, especially when it comes to handling trauma. After all she's seen and studied, she offers this: "Talk about it! Normalize and destigmatize mental health care. Mental health *is* physical health; the brain's a pretty significant organ that can malfunction just like any other body part. No one would shame you for not being able to heal a fracture with positive thinking, or not being able to survive type 1 diabetes with subcutaneous water, or halt a heart attack with some fresh air. Don't let anyone shame you for seeking medical care for your brain."

AN UBER DRIVER

Jamie and his documentarian team met Roger, an Uber driver, while being driven from site to site in Chicago. When he heard that they were working on a project that involved people with post-traumatic stress, he had a lot to say.

Roger has been driving an Uber for approximately five years. "People tend to open up to me. I've had quite a few conversations with those dealing with everyday issues in life, stuff that is bothering them, problems they are having with parents or spouses or people who are depressed. I've had a lot of interesting conversations. I've really connected with them. The main reason is because I've been through all kinds of therapy myself. I believe I've helped some of these people—they just need someone to listen, someone who is not going to judge them."

Before Ubering, Roger worked in sales for Comcast for nineteen years, until they downsized the staff in 2017. He got his CDL license and drove a truck for a short time but found that being in a small truck cab for three weeks at a time exacerbated his anxiety issues. "I'd literally have breakdowns at the truck stops. So here I am."

When Roger was young, he was heavily into boxing, as his family had been for generations. He even trained for a spot at the Olympics but eventually put that dream aside. "For the longest time, I worked as a sparring partner. A punching bag. I kinda wasn't allowed to unload, only take the punches. It got tiring after a while, getting beat up all the time. But boxing runs in the family. How everybody else gets together for a baseball game or a football game, well, we got together for boxing. I love the sport."

But boxing has its dark side, especially in the family home. "When I was a child, it was kind of rough growing up. My dad was very sick, but I didn't know it. He suffered from depression and oftentimes, especially when he'd drink, he'd take it out on my mom. He was a great, great guy, but he had a violent streak. She was at the end of a lot of beatings. Keep in mind, my dad was a boxer. He'd work out every day. He hit her with his fist. Me being a kid, I couldn't do much about it.

"One of my biggest regrets was not doing more for my mother. As the years went by, the beatings seemed to have subsided a little bit. But there were still times I'd get home and something had happened. As we got older, [my dad must have] realized he couldn't do that stuff around us anymore. When he'd drink, that was the worst part. I was terrified. I knew when he got home, my mom would catch a beating.

"I was definitely affected by what my father did. I think about it often. It does sadden me because I could have done more to help my mom. It's gotten easier over the years, but for the longest time I'd think about, it would upset me; I'd break down and cry. My relationship with my dad was a love-hate relationship. He could be funny and loving, but he was a very scary guy when he was drinking and going through his moods. There were many times,

I'll be honest, I wanted him dead, wanted him gone. I wanted my mom to live a normal life. I knew that was impossible as long as he was around."

Roger grew up in Chicago. Struggling to control his own anger and despair, he was getting into more and more trouble, joining a gang at a very young age. The members became his friends, even serving as father figures. They were violent, teaching him to channel his anger into street fights, but they were loyal to him and available in a way his father wasn't. His parents moved out to the suburbs to get him away from the wrong influences, but he kept going back to see his friends. Kept getting in trouble.

"I participated in a lot of gangbanging. You name it, I did it. Shootings, stabbings, beatings. I was a bad kid." When asked about an incident he regrets, Roger tells the story of when he and his crew happened across a guy from an allied gang who had been severely beaten. Roger grabbed the man's shotgun and went to the rival gang's house for retaliation. As he tells the story, his voice chokes up and he is unable to make eye contact. "Out in the street, I opened fire into the crowd. All I heard was screams. They had kids out there."

Those memories haunt the Uber driver. "The only reason I'm still alive is because of [my two daughters]. I've tried hanging myself. Shooting myself. At times, I've held a razor in my hands and just wanted to slit my wrists or my throat. An older gentleman in group therapy told me, 'If you was to do this to yourself, your girls are going to find you. No one else! You don't want to put them through that.' I believe that is one of the reasons I am still alive today. My little girls. I can't put them through that. I can't leave them to fend for themselves."

"While I'm broken, I think that I do have a gift to help others. But sadly it's very difficult to help myself. Ever since I was a child, I've experienced anxiety, depression, I'd break down, I mean really break down. I don't know what living a normal life is like. I'm hoping I get to experience that. There's so many things I wanna do. Like everyone else, I have dreams. I want to be successful. I'd have [careers or jobs set up] but I'd hit a barrier, a mental barrier. I just want to be happy, I guess."

AN ANIMATOR

Eduard, an animator, left his Romanian homeland in his late teens to get away from Catholic state rule. Once settled in the UK, he realized his anxiety was becoming a problem. He spent an inordinate amount of time seeking out doctors and any modality that would relieve the anxiety, which continued to worsen—he became "caged" in his apartment, anxiety making it impossible for him to leave. He found that the only way to force himself through the door was if he was inebriated. "The one day I went to work sober, my coworkers all asked me what was wrong with me."

A self-proclaimed disenchanted atheist, he says he wanted to believe it was possible to lead a normal life, but none of the medical interventions he tried worked. "I lost all hope," he says somberly.

Until he read about an American doctor and his dual sympathetic reset.

Simply knowing that something like Eugene's procedure might exist in the world "lit a fire in me again." He said he "was keen to get the procedure" and then to share his experience on the many social media platforms he had joined over the years around the topics of anxiety and depression.

He said it reminded him of an experiment he had watched on YouTube about the power of hope. It involved drowning rats, which he acknowledged was extremely disturbing, but the point of the experiment was that the rats who were saved from a drowning had then fought harder to live during a second drowning attempt. They had a sense of hope, knowing they'd already survived the traumatic experience once. For Eduard, that was an impactful lesson—hope of rescue carries massive power.[90]

No one in London seemed to know about Eugene's procedure in relation to relieving trauma, so Eduard decided to reach out to Dr. Lipov directly. Eugene was able to direct a British physician through the procedure, advising him to do the injections on both sides of the neck.

In an initial interview with Eduard, conducted over an international Zoom call, the animator was sitting slumped in a dark room, unable to make eye contact, not even with a camera lens. Then, he had the injections. His life was transformed. He is able to once again conduct a normal, daily life in or outside of his apartment without self-medicating. In a follow-up interview, he sat upright in a fully lit room, looked directly into the camera, and smiled. His eyes sparkled; his countenance was confident and calm.

Now, knowing firsthand how powerful this procedure has been for him, he is determined to share how he is living in a pre-trauma state with as many trauma sufferers as possible.

A BLUE-COLLAR WORKER

Rawley had warned his wife, Linda, not to come into the bedroom while he was sleeping. She'd always followed that advice, but one morning she needed to retrieve an item, so she tiptoed in, careful not to make a sound.

Neither of them is sure what woke him up—perhaps a floorboard creaked.

Rawley sprung out of bed, having felt someone moving through the dark room. What happened next is fuzzy in his mind. As the three-hundred-pound former trucker grabbed his wife by the throat, she started screaming. He heard a female voice yelling, "It's me! It's me!" but he couldn't place the voice.

Rawley pushed the woman through the hollow-core bedroom door while she screamed in pain and struggled to break free of his grip. He suddenly snapped out of his fog. The whole scene seemed unreal and might have been a dream, until Rawley turned on the light and saw the red rings his large hands had made around her neck.

Linda understood he didn't intend to hurt her, though she occasionally threatened to "kill him" if he didn't get help. She knew he'd been drafted to serve in the Army Medical Corps in the Vietnam War and how, at age eighteen, he'd flown from country to country, hauling bodies during the Tet Offensive, the bloodiest siege in Vietnam. During his first year back, he got himself hired—and fired—from thirteen jobs. He battled anxiety, irrational and constant anger, and depression.

"I often felt like I wanted to kill someone. I didn't do a good job of hiding it."

The Veterans Affairs doctors diagnosed him with PTSD. They did not yet know the biological reason for post-traumatic stress, that Rawley's sympathetic nervous system was stuck in the "on" position. While others felt safe driving a car around town or walking through a shopping mall or down a city street, Rawley was hyperaroused and hypervigilant, responding irrationally to simple events like a car backfiring with rage and violence. Or being awakened in the middle of the night by his wife.

Rawley's doctors embraced the gold-standard medical approach for post-traumatic stress at the time, prescribing anti-depressants for him. He'd dutifully taken his pills over the years, but unfortunately, they made his condition worse.

"All they did was fire me up and make me madder," he says.

Rawley's suffering became so intense that in 2010, in a near-suicidal state, he showed up at an event in Chicago close to his home in the city's northwest suburbs, where Dr. Lipov was speaking about post-traumatic stress. The doctor was launching his first nonprofit clinic in order to deliver his adapted SGB procedure (pre-DSR) to civilians and soldiers alike.

Rawley, deep in suicidal ideation, found Eugene at the event and told him he was having vivid nightmares that made it hard to sleep and was suffering from extreme anger and overreactivity to minimal stimuli. The doctor called his staff together and met Rawley at his clinic. His post-traumatic score was so high, Eugene scheduled him to have the SGB the next day—it was clear he needed fast treatment.

"It was only afterward that Rawley told me he was going to kill himself in my parking lot if I had decided against treating him," says the pain physician.

When Rawley arrived back at the clinic, he was so anxious he could barely cross the threshold. The staff was able to calm him down enough to insert an IV and sedate him. Then they moved him to a special X-ray table in the operating room (the use of an ultrasound came a year or two later). An anesthesiologist gave Rawley a sleep medication called propofol.

With Rawley asleep, Eugene's team cleaned his neck with an antiseptic solution. After using the X-ray to determine the correct insertion point on the right side of the neck, a 22-gauge needle was inserted into the skin and used to inject the anesthetic.

The entire procedure was completed in less than ten minutes. Fifteen minutes later, Rawley awoke with a droopy eyelid and red eye on his right side. These were signs that meant the sympathetic nervous system was anesthetized properly; both symptoms wore off within a few hours.

Rawley was visibly relaxed and calm after the procedure—a dramatic change. He reported later that he went out to dinner that night and was able to sit with his back to a door for the first time in forty years. When a waiter dropped a stack of dishes, he did not react.

The team followed up to make sure he continued to experience healing of his sympathetic nervous system, a determination made based on a reduction in his symptoms. For some patients, Eugene needed to administer the procedure two or three times, which was found to be true in Rawley's case.

Given Rawley's long history of PTSI, Eugene recommended he continue with cognitive therapy as part of an ongoing treatment plan. When someone has lived with post-traumatic stress for a lifetime, a sudden change in their condition, even a very positive one, may require emotional adjustments—both for the patient and for their loved ones, whose lives likely revolved around coping with the patient's post-traumatic stress symptoms.

A CEO

Donald was diagnosed with severe PTSI years after living through a horrific natural disaster.

He grew up with a twin brother and younger sister in a nice, middle-class neighborhood in Long Island, where he went to a good school and had solid relationships with friends and family. In his twenties, Donald was working for his father's business in the flavoring industry when he was offered a position running an operation for a large corporation out of Los Angeles. He worked hard and eventually opened a successful consulting business.

Soon after, on the day after Martin Luther King Jr.'s birthday—January 18, 1994—the infamous 6.7 Northridge earthquake rocked Southern California, killing at least seventy-two people.

In a 2021 interview with Jamie's documentary team, Donald recalls waking up that morning in Santa Monica. "At 4:27 AM . . . I got up. I felt that there was gonna be an earthquake . . . I went between my door frames and held on because I remember somebody telling me to do that . . . then the big jolt hit . . . I'm talking about an energy, a force that was so beyond anything I had ever felt. Literally, my body just levitated when it hit. It was so much g-forces that I'm five foot ten and the door frame is, say, eight feet, and I actually hit my head on the top of the door frame. I don't know even how I held on, to be honest with you, that's how forceful it was . . . Most earthquakes are rolling, so they tend to be horizontal. This was unique because it was actually vertical."

He watched in shock as the room "disintegrated" around him. A giant, three-hundred-pound slab of wood splintered off a case, slid across the bed where his head would have been, and shot out the window. "I didn't know when [the shaking] was gonna stop because it almost went on for what felt like a half hour, but it wasn't, it was probably a matter of, I don't know, twenty, maybe forty seconds or something like that. But time is relative when you're in this fear state, it's just, you don't really have any sense of how long you're in it . . . When it stopped, I was walking in shards of glass. I realized I had to get my sneakers on and get some clothes on because I had to try to get out of the house. And I go downstairs to open my door and I can't open my door because the door has fused because of the vertical energy."

Donald described how many of the cement pillars on the LA freeways turned into cement "mushrooms" because most of the engineering at the time had been designed around rolling quakes rather than a huge vertical energy transfer. When the Northridge earthquake hit, the weight crushed the giant cement pillars. He said, "That's basically what the earthquake did to most of the properties, that kind of damage. But what it did to me was more emotional damage."

His neighborhood was one of the few in Santa Monica to be red-tagged. His home and most of his belongings had been destroyed. When his father came out from the East Coast, Donald was surprised when the "tough guy" saw the devastation and began to cry—it was clear that his son was lucky to be alive.

"Life changed a lot for me after the earthquake," said Donald. "I didn't realize it at the time . . . If you've suffered [adversity], you tend to figure out ways of . . . dealing with it. Some people go through compartmentalizing their emotions, some people adapt to physical pain differently, things of that nature . . . I didn't realize that I had PTSD at that time, because in 1994, unless you were somebody who was in the military or something like that . . . they didn't really look at the average person going through an earthquake as somebody who might have suffered something traumatic that would cause PTSD. That it would have a huge impact on their life.

"So, I didn't realize [I had PTSD] at the time. But I knew I couldn't sleep. Every morning after the earthquake, I would actually get up at four thirty. I still do. It was literally like an alarm clock and my heart rate would go up. I was panicking and I would actually think the room was shaking. For the first year, I was a complete mess. I really was a complete mess. I had managed; I was starting my business. I had lost my house . . . But what really was the challenge was just trying to get a full night's sleep and the impact and the effects of that lack of sleep—what it does to you cumulatively is really pretty amazing. I tried a lot of things. I wound up getting hooked on sleeping pills, they were . . . I think they were Ambien and they had some weird side effects. And so I got off those.

"Finally, I didn't really feel the room shaking anymore [when I woke up]. Then, it was about a year or so after the earthquake . . . I went to Santiago, Chile, on business. I will never forget this: I'm standing at the Sheraton, and I had literally just started kinda sleeping better, and all of a sudden, the rooms are shaking. I'm like, 'Oh, God, you're so sick now, you just can't get over this.' I'm thinking this is all my imagination. I turn on the TV and I put on CNN and one of the newscaster says, 'We're just reporting that there's been a 7.2 earthquake in Santiago, Chile.' Oh my God, talk about luck. Needless to say, that didn't help my situation.

"I wound up continuing to have this sleeping problem. I've been living with it for years, and I feel it's changed my personality in a lot of ways. I take a half milligram when I start a day and every night to go to sleep. If I don't take that, I don't really sleep very well and I still wake up at four thirty. And some days, I still have the panic a little bit. I feel my heart rate racing. It's been that way since 1994.

"If the earthquake didn't happen, I wouldn't have been addicted to sleeping pills. That's number one. I don't know whether or not that has made it more challenging for me, to have the patience that I would normally have. I think when you have sleep disorders, I think you can be easily very irritable sometimes. Even though I really work hard to try to avoid that. Try to be patient and such. But I think your brain rewires. Tries to manage, whether it's a pain issue, or it's an emotional health issue, or whatever it is you're dealing with. You figure out a pathway around it. For me, I've often wondered how things would have changed. Whether it's impacted my decision-making process, things of that nature. [In] my relationships, I think there's a lot of issues that would have changed . . . And after a while, you realize that you don't wanna just keep doing the drug or the medication or whatever . . . but you're not able to fully manage it. There's certain things that are underlying that prevent you. So for me, it's always been like the elephant in the room.

"I'd like to get rid of it, but I don't know how."

Jamie was instrumental in making that happen for Donald. He says, "Within a week he was off Ambien, and he's now sleeping for the first time in thirty years. It should be noted, his close family did not want him to get the procedure. They were scared by it. Thankfully, his GP and his cardiologist told him it was perfectly safe. He has told me the treatment has transformed his life."

THE ANSWER IS HERE AND AVAILABLE

You've read the stories. You've seen the research and clinical data. There is help out there for Donald. And for you.

CHAPTER TEN

The Zeitgeist

Eugene discussed post-traumatic stress impacts and his procedure before Congress. He has been mentioned in dozens of high-profile newspapers, magazines, television programs, and NPR and BBS talk radio, and his procedure has been addressed in hundreds of medical journal articles. The US Army has done a major study that shows the procedure's high rate of efficacy and safety, and NYU has launched another massive study using functional fMRI scans to see the impact of DSR on the brains of those suffering from PTSI. The investment group and people behind Stella have worked aggressively to set up clinics with trained DSR doctors.

So, why haven't you heard of the dual sympathetic reset before now?

A PROBLEM OF CREDULITY

SGB: the acronym is immediately familiar to those in the medical field who regularly use a stellate ganglion block for pain management.

Yet, mainstream mental health circles remain uninformed of the procedure and what it can do for their trauma patients. That is changing, slowly. The research has been hyperfocused within military populations rather than diversifying into other traumatized populations, so mainstream clinicians aren't coming across this breakthrough treatment—the keyword searches aren't there yet. This is even more true of the updated nomenclature, *dual sympathetic reset* or DSR, as the most recent innovations (dual injections on both sides, needle placement using modern guidance systems, and refined injection techniques) on the SGB are still fairly new.

Another issue centers around believability. "One of my friends is a professor and he says I have a credulity problem," states Eugene. "He says it's hard to believe that I could have figured this out when no one else has, not since it was first used in 1925. But the reason I was able to do it was because I was motivated by my mother's death *and* I was researching hot flashes—it all came together because of those two things."

There is a great deal of credulity from both the medical community and the general public. Both are skeptical that a fifteen-minute procedure can reverse the symptoms of PTSI after years of therapy and carefully crafted pharmaceuticals have had limited benefits. *The appearance of an easy button seems way too good to be true*. Yet, look at the data. The substantial benefits and minimal harms are neatly laid out.

Finally, there is the pervasive false perception of trauma we discussed in earlier chapters. First, we rarely see ourselves as having suffered to the same severity of frontline workers or sexual abuse victims, believing our trauma is not as important or as "big." Second, self-reliance and normality are idols in the American psyche, while trauma causing post-traumatic stress is still believed to be rare. That is because the stark disparity between how we think we should "be" causes us to feel like we must minimize or completely sublimate how we really feel or are reacting to stress, in order to save face in public or even within ourselves.

Because those who *are* traumatized are considered rare, and that level of trauma is "not normal," we often keep our own pain hidden. Even people who are clinically diagnosed with post-traumatic stress and receiving treatment will sometimes feel shame and keep it a secret. This keeps knowledge of workable treatments, including DSR, from being passed along via word of mouth to the average person on the street—and, again, we believe at least 40 percent of the population at large has PTSI.

If you do happen to hear of Eugene and his procedure through word of mouth and perform a Google search out of curiosity, you will be bombarded with information—mostly sound, data-driven responses and articles, like the 2021 write-up in *Forbes* magazine.

But you will also find criticism, much of which is skewed or out of date. So, how does a layman parse out what is fact based and what is not?

ANSWERING THE CRITICS

Eugene Lipov, MD, has had to defend himself and his procedure many times. "As it should be," he states. "This is medicine. This is science. It is right to seek answers."

But the 2005 *Chicago Tribune* hit piece that was published when he first discovered the revelatory use of SGB on hot flash patients still rankles him.[91] "No one talked to me. And the 'experts' they quoted did so without follow-up. Further, they treated the woman they quoted as if she were stupid for taking the 'risk'—though they knew there was minimal risk, since the stellate ganglion block had been used safely for a century. One of the 'experts' went so far to say he didn't believe the results were real, that it was a 'crock,' and that these women should try yoga and drinking cold water to curb their extreme and persistent hot flashes. Yoga!"

The chief of gynecology at Northwestern Memorial Hospital at the time was one of those interviewed in that article, supplying a snide sound bite, saying, "Whenever we have a new procedure, one doesn't do it casually or cavalierly. One should have evidence of why and how long it works."

This stung Eugene. He knew he had found something safe and effective, and yet the medical establishment went out of their way to make him look reckless. And why? As long as he continued with caution and there were nothing but positive outcomes, how come his peers weren't willing to keep an open mind, to do what was best for their patients? "What if no one had tested penicillin? And that *was* an unknown. This is not. Neither the method nor the local anesthesia that is used in the injection are new."

The harm comes from withholding this safe treatment while people are needlessly suffering. And, ironically, years after the *Tribune* article, Northwestern received a $4 million grant from the National Institutes of Health to study the role of SGB in treating hot flashes *due to its successful track record.*

Luckily, this type of sensationalized press only made the pain physician double down. He would continue to research and perform clinical trials and bring in other professionals and fight the good fight. The lives of people were on the line.

Since then, he has had much, much more good press than bad, but he is careful not to ignore either.

WHEN THE BIG DOGS BARK

In 2007, US Senator Barack Obama (a year before he was voted in as the forty-fourth president), wrote a letter to the army in support of Eugene's procedure.

Senator Obama wrote, "There is a growing body of evidence to suggest that PTSD is afflicting a growing number of our heroic service members . . . [It] is important to consider any new approaches that may hold potential for helping our service members get the care they need." The request was denied, mainly due to the small number of SGB (not yet DSR) patients up to that point, but that didn't stop Eugene from funding his own limited clinical trial, just to keep the ball rolling.

An article in *Wired* magazine came out in 2010 titled "Obama Loves This Freaky PTSD Treatment; the Pentagon, Not So Much." The author tried for balance and an open mind, but one thing that sticks with both Eugene and Jamie Mustard now is the statement: "Of course, there's the undeniable fact that SGB injections are a Band-Aid treatment, rather than prevention or all-out cure. But according to Eugene, they're the best we can do."

Eugene *never* felt SGB was simply a Band-Aid, and years of results have proven him out.

In 2010, Eugene spoke to the House Committee on Veteran Affairs, seeking support for SGB as a treatment, to no avail, and then he spoke to Congress. Audio is available at www.dreugenelipov.com.

In 2011, Eugene approached DARPA for research funding. The Defense Advanced Research Projects Agency, established in 1958, is an agency within the Department of Defense (DOD) responsible for catalyzing the development of technologies that maintain and advance the capabilities and technical superiority of the US military.

DARPA said no.

When asked by a general outside the meeting about the outcome, Dr. Lipov said, "Well, thirty thousand troops just died with that decision." He wasn't giving in to histrionics. Since that day in 2011, eleven years have passed and roughly 88,000 people have killed themselves—Eugene believes a significant number of those people could have been saved if they'd had access to DSR.

THE NAVY STUDY

The first randomized, blind sham-controlled study for the use of SGB (pre-DSR) in post-traumatic stress treatment was done by the United States Navy in 2014. Steven R. Hanling, MD, and colleagues, from the Department of Anesthesia and Pain

Medicine, Naval Medical Center in San Diego, summarized the results in a 2015 article, in which Hanling and the other authors suggested that Lipov's procedure lacked efficacy, and "although previous case series have suggested SGB offers an effective intervention for PTSD, the study did not demonstrate any appreciable difference between SGB and sham-treatment on psychological or pain outcomes."[92]

However, the VA printed a rebuke not only of the navy's findings but Dr. Hanlings's methodology. After a thorough review of the study, the Veteran Affairs Evidence-based Synthesis Program (ESP) reported, "Because these findings come from a single study with imprecise findings, moderate methodological limitations, and did not directly focus on clinically relevant outcomes or use the most common administration techniques, they provide an insufficient basis upon which to draw conclusions about SGB for treatment of PTSD in Veterans."

Specific findings from ESP regarding the study headed by Hanling included:

"Although in previous case series the most commonly used anesthetic type and dosage used have been 7 cc of ropivacaine, or bupivacaine 0.5% solution, this trial used 5 cc ropivacaine, a 28 percent lower dose, and provided no rationale for doing so. Also, it is unclear whether the ropivacaine injection actually reached the stellate ganglion in all patients. Although the stellate ganglion is typically located at C6 to C7, the level of target needle placement was C5 to C6 in this study. Although the study author confirmed that the injection was 'typically' at C6, some could have been at C5 and missed the stellate ganglion.

"Other commonly criticized limitations of this trial that may have altered SGB effects include that (1) it used an inappropriate population who were in the process of disability evaluation and may have had secondary financial incentives to resist treatment, and (2) the use of saline instead of an active control that mimicked the side effects of SGB was potentially inadequate and may have reduced the effectiveness of the blinding, as patients may have been able to easily tell if they received SGB or sham based on the occurrence of the Horner's syndrome eye droop."[93]

Eugene did not have a problem with negative feedback if justified, but the inaccurate portrayal of the use of SGB and the biased review of results was too much, including questioning the diagnostics and placement of the injections; Eugene published a rebuttal in the publication that first published the flawed study.[94]

There are few risks with the stellate ganglion block, which has been used for almost a hundred years with minimal effects reported. As referred to in chapter one, a report in 1992 (before SGB used ultrasound or fluoroscopic guidance for even more precise placement) studied the results of 45,000 blocks—the incidence of severe complication was 1.7 in 1,000, with no lasting effects and no fatalities. The complications were due to puncturing a lung or hitting a key artery, or convulsions related to the central nervous system toxicity from rapid anesthetic absorption. At the time of the study, Eugene knew his complication rate was even lower, due to his targeting the C6 vertebrae rather than the traditional SGB C7, which is closer to arteries and the lungs.

THE ARMY STEPS UP

Not long after Dr. Lipov's "Letter to the Editor" was published, a rigorous, three-site study on SGB was concluded. The study, which was funded by the Pentagon and ran from May 2016 through March 2018 in three Army Interdisciplinary Pain Management Centers, supports Eugene's claims of high efficacy with few side effects.[95]

In 2021, award-winning director and producer Michael Gier released *Wounded Heroes,*[96] a documentary focused on successful alternative treatments for PTSI. Gier included a substantial interview with Dr. Ali Turabi, one of the doctors and principal investigators of the three-site army study.

The study, based on active-duty soldiers with clinically identified PTSD who underwent SGB, was funded by a grant from the National Institutes of Health. A control group and the experiment group were studied by military doctors in Landstuhl, Germany (where Turabi was based at the time), Fort Bragg's Womack Army Hospital, and Tripler Army Medical Center in Hawaii. Active-duty physicians performed the procedures with no extra incentives for their involvement—an important factor in the outcome, as the study wasn't funded by big pharmaceutical companies focused on certain outcomes.

The treatments and accompanying research were done over a two-year period of time, during which the physicians were earnestly seeking valid, clear information on what worked or didn't work for their patients.

In the documentary, Turabi reveals that each patient had first one injection and then another injection two weeks later; the results were followed up at eight weeks. Every participant who received the stellate ganglion block showed significant improvement in their symptoms: the group who received SGB was observed to have twice the amount of relief from PTSI symptoms as the placebo group. The placebo group unknowingly had an injection of saline placed directly into the muscle tissue just below the nerve group. According to Turabi, some patients can experience a placebo effect in these kind of studies, and "the patients do experience a benefit, even though you didn't do the actual procedure. But in this study, they did have a benefit with the stellate ganglion block and they *didn't* have a benefit for the placebo. So we know that it does work."

Turabi acknowledges he initially was not interested in SGB, not sure if he believed in the validity at all, but the undeniable results changed his mind.

"I was extremely skeptical. I'm an anesthesiologist pain physician—I don't necessarily even treat patients with PTSD. So, when this came across my desk, I thought: *A shot for PTSD? I don't think so.* But after going over the theory behind the procedure and discussing it with some of my colleagues, I decided it was worth a try. I think I was very skeptical, very objective about it, throughout the process. I was like, let me see what happens. But again, the results speak for themselves."

He adds, "But even before the official results came out, I knew that this was something that worked because of my patients' responses and their spouses' responses. Patients were giving me hugs, spouses giving me hugs, telling me, 'My life is totally different. My husband's life is totally different. Thank you so much.'"

Once Turabi and colleagues completed their bigger, more comprehensive study and shared the results, one of the authors of the navy study, Dr. Robert McLay, once a naysayer, now claims to feel confident in the validity and benefits. In an interview with Gier, he says he intends to use stellate ganglion blocks more widely.

"We did a bigger study," said Turabi. "We use a gold standard for measuring PTSD, which is called the CAPS five score.

Those patients did well, our results are there, and it's clear that they do get better."

Eugene was also interviewed by Gier for the documentary. On camera, he stands by his procedure. "The risks are extremely low because we use ultrasound or some other guidance to make sure we're in the right place. But it is still a medical procedure; it's an injection. There are risks any time you have any type of procedure. That's why we closely observe people here."

Also, SGB will result in Horner's syndrome, which is expected and dissipates within eight hours. "Basically, you may have some nasal stuffiness, warm hand, and redness of the eye, and the droopy eye. Those are some things that we expect to happen after a stellate ganglion block. And they only last for about six to eight hours and then go away. There are some exotic things that could happen. Irritation of some nerves, some people may have hoarseness, they may have persistent eye droop . . . extremely rare.

"I've never had any persistent complications occur with any of my patients. If done correctly, the ultrasound technology is so advanced that we can see all the important structures in your neck and avoid them. We can see right where the needle is going. And so those things all really help with, with, uh, you know, decreasing the complication rate."

Turabi, the one-time SGB cynic, scoffs at critics who tried to float the idea that the results were paid for. "There's no big moneymaking scheme here!" he says. "There's no big pharma involved. It's done in military facilities by military doctors. We didn't make anything extra off the study. The procedure itself costs two dollars for the medication, so big pharma is never going to fund the study for this, because they're not going to make any money off of it. So, I think in fact, it's the opposite. It's done in military facilities and we had no incentive to do the procedure—beyond the fact that we wanted to help people."

Gier asks a number of pointed, direct questions. His goal is to find the truth behind treatments, not to just blow smoke. He tells Turabi that one critic has said, "SGB is a medical treatment that is trying to solve a nonmedical issue."

Turabi replied, "My argument for that is, how do you know it's a nonmedical issue? You know, we don't really know that much about PTSD. I think it is a medical issue. You've got your sympathetic nervous system that is overactive and that's a medical issue and it's causing your PTSD: anxiety, hypervigilance, sleep issues."

Gier then comes back with a question that has been brought up many times. "I have another critic, and they're saying if you prevent the reaction of fight and flight from happening with an injection, then the owner of the brain is prevented from having that natural reaction that they should have. So, is the shot taking away that natural reaction?"

Dr. Turabi has experienced the same responses as Eugene. "I've actually had a lot of Special Operators come in and say, 'I *need* that fight-or-flight response. I need to be able to kick down a door and look around and be vigilant or hypervigilant and see what's going on.' But this [treatment] doesn't take it away permanently. This takes it away for six hours.

"Now, the guy who is kicking down the door, he doesn't need that response all the time. When he's sitting and watching a movie, relaxing at home, he doesn't need to be hypervigilant, right? He actually needs the opposite. He needs to be able to relax, calm down. That fight-or-flight response should be there only when it's needed.

"There was a study done where they tested patients after they had the stellate ganglion block, their ability to respond quickly—and [their responses] actually improved after the study. They had a better ability to respond to situations where they needed that fight-or-flight response.

"Most of the referrals that I got for this procedure are from psychologists and psychiatrists. A lot of them have reached a wall with the type of treatments they can provide, and they send the patient to me so that they can get the block to provide them a window to sort of do these other treatment options. And I think that those treatments are more effective during that period of time, after they've had the stellate ganglion block."

MAINSTREAM MEDIA GETS IN THE GAME

Some mainstream media outlets *have* covered the benefits of DSR performed by Eugene and other physicians. As more and more patients speak out about their experiences with the procedure—like Matt Farwell in his 2016 *Playboy* article or Medal of Honor recipient Dakota Meyer on a 2019 Joe Rogan podcast—the list will continue to grow.

Here is some of the most prominent coverage:

- *The Doctors*, a spinoff from Dr. Phil's medical talk show, is in the fourteenth season and is regularly listed among the top syndicated talk shows. A team of medical professionals (and sometimes celebrities) discuss health-related topics, and they are known for taking questions from viewers too embarrassed to bring up certain issues with their own doctors. Eugene first appeared on the program in November 2014, then November 2016, and again in May 2021, discussing his procedure and how successful it has been with both hot flashes and post-traumatic stress. The May segment, "PTSD Treatment Helped Two Men Heal from the Trauma of Gun Violence," is described by the network as follows: "How a PTSD injection works to give people their life back! After James heard about his friend Derrick's success with anesthesiologist Dr. Eugene Lipov's PTSD treatment, he decided to go ahead and do it. James has been shot five times in the head, and he hopes to be able to sleep again at night. Derrick and James join *The Doctors* with Dr. Lipov, one week after James received his treatment."[97]

- *Sixty Minutes* on CBS ran a segment called "New Army-Funded Research Shows Promise for PTSD Treatment" in November 2019. Earlier that year, they ran three similar segments. The episodes highlighted the many benefits of the treatment, interviewed multiple patients, and even followed a trial participant before, during, and after SGB.[98]
- *People* magazine covered the procedure in May 2020, *Forbes* magazine in February 2021, and the *Wall Street Journal* in 2017 and then again in 2019. All three of the journalists covering the story started with a critical eye but ended up touting the benefits and encouraging readers to look into what is now being described as a miracle procedure.

BIGGER AND BETTER

Professionals in the medical world are buying in. Behavioral health specialists—psychologists, licensed clinical social workers, psychiatric nurse practitioners, and psychiatrists—with expertise in the treatment of PTSI participated in a 2020 qualitative survey analyzing their personal experience with pairing SGB with trauma-focused psychotherapy. As a result, SGB was rated as "useful" as those therapies listed as "most valuable" by the American Psychological Association Clinical Practice Guideline for the Treatment of Post-traumatic Stress Disorder. When asked which trauma symptom SGB most impacted, 96 percent of respondents identified "Arousal/Reactivity" with an overall 95 percent affirming they would recommend SGB to colleagues.[99]

And the biggest, most strenuous study to date is now in the beginning stages, as this book goes to press. NYU has approved a $3.8 million clinical trial focusing on DSR and the benefits, which began recruiting patients in April 2022. They will be using a functional MRI on patients before and after the procedure to measure improvement of the patient's brain function, especially the amygdala.

"It's quite extraordinary! I'm humbled. It's finally happening in a big way. And it's only happening because of generous donors—Linda Vester and Glenn Greenberg. Linda has gone through the procedure herself and knows exactly what a game changer it is," says Eugene.

Glenn and Linda of the Glenn Greenberg and Linda Vester Foundation have been generous with both their time and money in regards to getting the NYU study off the ground; they have been heavily involved with working with trauma patients for years and are invested in better healthcare for those who suffer from PTSI.

"We hope that the study will have a definitive answer about the efficacy of DSR," states Eugene. "The NYU team is working to understand some of the mechanisms behind the DSR effect on PTSI. If the results of a study of this size done at such a high-quality institution demonstrates the anticipated results, we believe this will finally lead to a wide acceptance of DSR for a treatment for PTSI."

WHY IT'S WORTH IT

According to Eugene, individuals should hold on to hope that PTSD is treatable and that they may find rapid response and relief. PTSD can be controlled, and patients can go on to live a full life after treatment. Also, society must stop shaming people with mental conditions and start seeing them as people with true medical issues that need to be addressed.

Finally, he believes the government should provide grants for new research instead of pouring money into dated and ineffective treatments. The government should also ensure insurance carriers provide the needed care for people suffering from mental illness.

Here is just one of hundreds and hundreds of telling stories around the veracity of DSR:

Eugene recently treated a seventy-year-old Vietnam vet using DSR, and after the procedure the man began sobbing, unable to stop. The pain physician was stunned, concerned he had "broken the guy somehow." After calming down enough to speak, the man shared how he had not felt an emotion in over fifty years. He had been absolutely numb, unable to feel neither sorrow nor joy. Only minutes after the procedure, a rush of emotion swooped in. He was once again able to feel.

The only way to keep patients safe—and help them heal to the point that they can once again smile at the sun on their face or laugh at a joke—is for the medical community to continue applying the scientific method, earnest peer review, and critical thought. Using these methods, the local anesthetic and original procedure in the century-old stellate ganglion block have proven themselves safe and effective for blocking pain; Eugene's innovative use of SGB, developing it into the dual sympathetic reset, has also established itself as safe and effective for other uses. The world just needs to catch up.

CHAPTER ELEVEN

The Big Stories

Jamie was sifting through footage of documentary interviews, trying to decide what to include in the book.

"I hesitate to bring up these big f-ing trauma stories. I really, really do. Our goal here is for every individual to know that *all* trauma, big or small, can cause post-traumatic stress *injury*. That's it."

Stories from a famous foreign war correspondent or a decorated Delta Force operator come from people who've lived different lives than most of us. But in those lives, just like in your life or a family member's life, the trauma they've endured has caused their invisible machine to break down and need a reboot. While their stories *are* sensational, hopefully you will also see these people as leaders and follow their example when it comes to admitting to suffering from stress and trauma.

It was the famous journalist's daughter who finally forced the correspondent to face that truth—her girl has suffered from traumatic brain injury due to multiple concussions and her own PTSI for quite some time. The Delta Force commander would not admit to a diagnosis of PTSI until his wife, while dealing with her own symptoms from past trauma and acquiring secondary PTSI because of his behavior, made him look in a mirror. Whether you are living a mundane existence or fighting terrorists, if you've suffered trauma, you can feel normal again. And if you're human, you've suffered trauma.

BEHIND THE CAMERA

A fancy car, a fancy house . . . none of that shields people from the bad stuff. Linda Vester acknowledges she is a wealthy white woman and that people who don't know her might perceive her as above the fray of life. But, like of all of us, she has experienced trauma. Her childhood was not easy, growing up with a suicidal mother and an alcoholic father in a house that never felt safe.

In an interview with Jamie, Linda talks about how, as an adult, she chose a career that put her in harm's way more often than not—a war correspondent. Even when in the States, being a reporter wasn't easy; she endured two incidents of sexual harassment from a high-profile news anchor in the nineties and then had to deal with the fallout when her accusations went public.

She is grateful for the life she has now. Recognizing the hurt in the world around her, she spends large amounts of her time and money to help those who remain in distress.

"We're all wounded souls in some way, shape, or form. You know, these days, people start to say, 'I'm more wounded than you are' . . . I don't look at it that way. I don't; I just see the wounds. I want to help people. But I can understand if you're coming from a place of pain or deprivation that you can look around to see somebody else who seems to have it better on the surface and think [how can they have] problems, they got money, what could be wrong? But I have yet to meet a single soul on this earth, no matter how rich or poor, who gets a free pass from trauma or crisis in their lives."

As well versed as she was in the nature of trauma, she quickly noticed when, at a dinner party in Nantucket, the Navy SEAL next to her was laughing easily and telling jokes—he was far too happy for a guy fresh off combat duty and the ravages of war. She smiled at him and said, "I'll have what you're having." The soldier told her about his visit to Eugene Lipov's clinic and the amazing procedure that had changed his life. Two weeks later, Linda had the procedure. Then, she went out of her way to make the acquaintance of Eugene.

Months later, sitting on a sofa in her living room with the documentary crew, she says, "I can't say I love peeling off my protective layers to go back in and talk about the trauma. And to have my daughter talk about hers, too . . . but I'm on a mission. I have been blessed with being able to get a treatment that should be available to so many people. I feel compelled to do this because it could set so many people free."

She pauses in the interview and then says emphatically, "This thing set *me* free."

Linda Vester, a reporter and news anchor for national news networks for twenty-plus years, covered stories from violent sex crimes performed by priests, to the Rwandan genocide, to the atrocities in Somalia.

"The [worst] image that comes to mind is from the Gulf War. Being in Kuwait City just as it was being liberated by American troops. A young man who spoke English found my cameraman and me, and he said, 'I need to show you where they tortured people' . . . He showed us the Iraqi torture chambers and he took us to the room where they kept the acid baths and showed us how the Iraqis had tortured people with acid.

"That's one of the things that is burnt into my memory. [Then] there's the Rwandan genocide—bodies slashed by the road. People dying left and right."

There was also her kidnapping while on assignment in Somalia. "I was reporting for NBC. My cameraman and I were in Mogadishu, the capital of Somalia, and we were covering the humanitarian disaster. People were starving and the militias had taken over the country and they were keeping all the food from the people, so they were starving to death . . . We were going to report on the situation at one of the hospitals. My cameraman and I left the NBC compound, driving our little pickup to the hospital in Mogadishu. [Halfway there], we got stopped by these militiamen, who they called 'technicals'—basically, these young fighters who were high out of their minds and armed to the teeth. They surrounded [us] and said we want your money and your camera gear. And my cameraman, who I adore, had a great sense of humor; they said to him, 'You must be a very rich man because you are very fat and you have a white wife.' He said, 'She's not my wife and you can have her if I can keep the camera gear.'"

Linda laughs ironically and rolls her eyes. "I was like, Gary, if we make it out of here, I'm going to kill you. But, fortunately, the Red Cross doctors at the hospital noticed we weren't showing up and they sent out their own armed guys, who found us and rescued us."

RESPONDING TO TRAUMA IN THE MOMENT

While she and the cameraman were being held, Linda recalls being struck by the slow passage of time. "All of a sudden, things were moving really slowly and I don't know what was going on inside my brain that made it perceive that, but there have been other times when I've been reporting, when things were life or death, I'd felt time slow down. So, I recognized that this was a pivotal moment that could end really badly. In some ways, when time is slowing down, part of my survival mechanism was for [my brain] to not allow me to recognize the gravity of the situation I was in. The survival instinct in me had me thinking, *Okay, can we joke with these guys? What can we do to get ourselves out of this?* It wasn't allowing me to perceive the magnitude of what *might* happen."

After returning from the Gulf War, Linda knew she was different. "At one point, I went to go visit my parents, to reassure them that I was okay, and we were having dinner and there was a sudden clap of thunder and I dove under the table. My dad, who had been previously an army doctor, looked under the table and said, 'Honey, I think you have PTSD.'"

She considers herself fortunate that her post-traumatic stress symptoms were "pretty minor" at the time. "It was really just a noise sensitivity from having been around bombing and so on, so I was hyperaroused, but able to get treated pretty quickly."

Her PTSI was much worse after covering the Rwandan genocide and then the refugee crisis. "I found it difficult to reinsert into American society. I had spent well over a month in a conflict zone where things were sort of life and death every day and people were starving to death and they were dying of horrible diseases in front of me. I was surrounded by corpses. I would come out of my tent in the morning and find dead bodies. So, I found it difficult to leave and come back to American society, where you go to the grocery store and you pick up detergent and go to the cash register and just go home."

Linda suffered from extreme anxiety, over-awareness, hair-trigger anger, terrible sleep, and an almost constant sense that something bad was about to happen. "Other symptoms involved avoidance behaviors. I had to avoid people or things or television shows or books or subjects that were triggering. People who were good friends, people I cared about, but they were associated with something painful and traumatic, so I couldn't be in touch with them because it's just too hard, too painful. I had nightmares that would come and go. I felt that my brain was stuck in hyperarousal mode . . . I would have these disproportionate reactions to situations where you would just maybe have normal anxiety or a little anxiety, but I was always at eleven. I couldn't modulate that, and I think that's just because my brain got stuck there."

RESOLUTION

"I was a person who has been blessed with a loving husband and healthy children, a good life, lots of friends . . . but I had to work at being happy because of all the things that had happened to me."

When talking about the dual sympathetic reset injections she received, she says, "Everybody deserves this freedom. And if that means I open up a little bit and share the muckity-muck stuff in my life then, sure, sign me up. I want to do this to help other people find out about this and hopefully get the treatment.

"I wish I had known about the procedure before I spent time on other treatments and modalities, because everything worked so much better after Dr. Lipov's treatment . . . I could come back and do everything else that life requires much better after I had the shot in my neck. I've tried different kinds of antidepressants but nothing nearly as effective as the Lipov procedure; I've tried MDR and TMS and neurofeedback and a bunch of other acronyms that I can't think of right now. And nothing was as effective as this shot in my neck. It took only fifteen minutes.

"For the first time since I was a kid, I had this kind of instant access to joy that I haven't had since I was like eight or nine. That I could just be happy at seeing a bunny rabbit in the yard. Just because. And I hadn't felt that in so long. When it happened, it was like, oh yeah, I remember this when I was a kid and it feels good!

"And I thought it was miraculous. I hadn't even thought to expect I might get *that*."

A CHILD IN PAIN

As Linda pushed open the door to leave the clinic, reveling in being symptom free, she knew she'd be back. Meredith, her child, needed Eugene and his miracle procedure.

Meredith, now nineteen, had begun having suicidal ideation and engaging in self-harm at the age of thirteen. Linda believes that being the victim of middle school bullies led Meredith to the depression crisis. But then a prescribed antidepression medication "sent her over the edge." Linda states, "She was self-harming.

She was suicidal. She was a mess. I was terrified I was gonna lose my child—I had eyes on her twenty-four hours a day and I was sleeping in her bed at night so that she couldn't get up and do something impulsive or dangerous without waking me. I felt that I needed to be there to save her life."

Meredith's teenage years were filled with angst and sorrow.

"When I was twelve, my family basically picked up and moved from this big bustling vibrant city [New York City] to this tiny little white suburban town," says Meredith in an interview with Jamie. "And I came from, you know, taxis and seeing every kind of person imaginable on the street and feeling like almost comfortably small and unnoticeable in the city . . . to now being surrounded by like Lilly Pulitzer [clothes] and boat shoes and crew socks and lacrosse and tennis and, you know, people driving around in Jeeps. And I was like, oh my God, I do not fit in here. If you didn't play a sport, you were nobody. And I did not play sports. I was a city kid.

"So I started playing lacrosse. I fell in love with lacrosse there, but . . . I stood out so much and I didn't really have the skills to make friends . . . It didn't feel good when I was feeling like I didn't have many friends, and that the [other middle schoolers] thought I was the weird girl from New York City who wears these bright red moon boots around everywhere. But I don't think it was really outside the normal scope of normal teasing in middle school.

"On a fundamental level, it's hard to feel alone . . . I felt isolated. I felt unwanted," she continues. "It made me withdraw. It made me feel so different."

The bullying lasted throughout seventh grade. Though she claims her experience probably wasn't much different than most others at that age, she refers to it as "the Great Unraveling of Meredith Greenberg." Her parents sent her to a private school, hoping that would help, but the girl continued to combat deep depression and trying to "fit in." A psychiatrist put her on Prozac, and she had what the pharmaceutical company claims to be rare side effects, inducing a bipolar state and manic episodes as well as auditory and visual hallucinations. She ended up in a psychiatric ward for a short time, to get the self-harm and suicidal ideation under control.

There was more at play in her life than the repercussions of bullying.

As mentioned, she had begun playing lacrosse in seventh grade. She soon discovered she loved it, playing year-round, six hours or more a week, even hiring a private coach. Meredith came to define herself as a serious athlete, dedicated to becoming a college lacrosse player.

"When I got my third concussion, it came to screeching halt. It was a bizarre experience. Having something be such a huge part of your life, and you're convinced that this is gonna be such a big part of your future."

When her neurologist told the tenth grader that she was not going to allow her to play lacrosse anymore, Meredith says, "It was like that moment in a movie when the record needle lifts. I asked her to repeat herself because I wasn't sure that I heard her correctly. And she was like, 'Yeah, after three concussions in the same sport, you know, I can't recommend that you keep playing contact sports. I'm gonna have to write a note to your school saying you can't play contact sports anymore.' I cried every day for the next week. I never thought I could have such an emotional connection to a sport, but it had become such a big part of my life that it felt like someone had just ripped off one of my limbs. I was missing a core part of myself. And to an extent, that feeling never really went away. I still wish I could be playing lacrosse. I identified myself as a lacrosse player, as a goalie. That's who I am . . . who I was."

Meredith was dealing with a high allostatic load already, due to the bullying. Then, the three concussions over a short period of years lead to traumatic brain injury, which affected her cognitive abilities. When she was forced out of the sport that had given her an identity, that was a tipping point for her allostatic load. If she hadn't already had post-traumatic stress, she did now. The teenager was suffering symptoms of both TBI and PTSI.

"I would come across a situation where I would feel anxious and then it was, you know, like a record player, just kind of skipping and playing the same thing over. I could not get myself off a certain thought, or out of this anxious feeling, and I would spiral. There were times something would happen during my day and it would just snowball into this 'thing' where it was two in the morning and I was sitting on my bathroom floor sobbing and I had no idea why. I just couldn't stop."

A MOTHER'S GRATITUDE

When Meredith's mother returned home from her appointment with Eugene at the Stella Center, she made a phone call informing Dr. Lipov that if he could treat Meredith in the same way he'd treat her, she would donate one million dollars to a study to ensure others would also have access to this treatment.

Linda refers to Eugene as a groundbreaker, a visionary, and a healer all in one.

"I have seen my daughter set free from paralyzing anxiety and ruminating thinking and being stuck in her head. To see her be able to just weather the normal kind of stuff that high school students have to weather."

And, like Linda, Meredith got her joy back.

After Meredith woke up from the procedure, Linda says her daughter "started riffing on all this hilarious stuff! She was just so funny! I said to her, 'What's got into you?'"

Meredith laughed at her mother and said, "I don't know. Well, I guess I'm hilarious now."

Her mother was overcome with emotion. "Seeing this kind of carefree joy again was priceless. It was just priceless. And to see her settle into it as each week passes, that she is just even more confident. Now, she feels like she *owns* herself. All the pieces fit now. And to see her own that, as she heads out of high school into college, is great. It's beyond great.

"But it also just makes me think how it was by the grace of God I was able to provide for her.

"I wish every kid who has been hurt could have access to that.

"And that every adult who has suffered can have access. How things would change if people could get this simple, safe treatment and be set free from anxiety and from trauma and PTSD. Oh my gosh. How things [would be] so different."

Since then, Meredith has followed up with brain scans at Dr. Amen's clinic and has reported successfully treating her TBI.

Also, the Glenn Greenberg and Linda Vester Foundation has donated to the major NYU study on DSR coming out soon, as described in the prior chapter, and they have endowed a psychiatry chair for the study of PTSD at Yale School of Medicine, and they support other causes for veterans.

Meredith Greenberg founded the Athlete Concussion Foundation in 2021,[100] and she is a board member on the Erase PTSD Now nonprofit.

THE ULTIMATE WARRIOR

In another round of interviews, Jamie spoke with Jen and Tom Satterly, a powerhouse husband-and-wife team who have established a nonprofit for warriors with post-traumatic stress.

Command Sgt. Major (retired) Tom Satterly was deployed to combat zones countless times and led hundreds of military missions. The recipient of sixty-four medals and ribbons, including a Silver Star and four Bronze Stars, one for Valorous Acts, he's put his life on the line for this country again and again. He's the guy you want covering your six.

Tom served in the army for twenty-five years, twenty of those in Delta Force, the military's most secretive and elite Special Operations unit. He retired while serving as the unit's command sergeant major. As an operator and CSM, he participated in or led many high-profile missions, including Operation Red Dawn (Saddam Hussein's capture) and the Battle of Mogadishu, the longest sustained firefight since Vietnam. The movie *Black Hawk Down* is based on the eighteen-hour firefight in a city filled with enemy combatants alongside innocent civilians, and the characters are portraits of Tom and his teammates running the Mogadishu Mile to safety.

Although the massive physical, mental, and emotional toll of war built up over the years, Tom refused to acknowledge his post-traumatic stress symptoms until the last few years.

"Post-traumatic stress? When it started for me, I didn't realize I had it." He shrugged, talking to Jamie in a 2021 interview. "[The stress] was a tool to do better, be better. It was a tool to never get in *that* position [again]. I was going to train, train, train so Somalia would never happen again. My focus, my energy, was all directed into work. I had zero empathy and compassion. I lost three wives." He chose to "keep his edge" and serve his country: each time one of his wives suggested he make a choice between his unit or his marriage, he was unwilling to walk away from years of sacrifice and dedication.

Then, with a heavy sigh, Tom discusses how the third marriage was different. "I had a son with her. Even though I felt differently about him, I was still gone all the time. My focus was still my work. I haven't seen my son in three to four years—I'm actually gonna see him today. He just didn't want anything to do with me. I was a mean person. I didn't focus or pay attention to him. You know, when I did come home, I was preparing to go away again. Mentally, I was already deployed. And then by the time he grew up and I was retiring . . . I wasn't in any shape to be around him anyway. That relationship was tanking and failing."

WHEN FEAR PURSUES BRAVERY

As Jamie listened to the stories, he came to think Tom Satterly was a guy who could do anything; the warrior was highly trained, motivated, full of testosterone, part of the top unit in the world.

"I was proud of what I could do," said Tom. "But, then, overseas, you see your friends get shot by children. You wake up to the reality that this isn't the Hollywood version of what a military person is. When you're outnumbered three thousand to one, it's heartbreaking when you can't come up with a way to get out. A convo trying to get in, helicopters trying to get out, and you're stuck. That's hopelessness. I lost the feeling of being impervious to death, and that fear of death drove me to work harder. I just put my head down and worked harder and harder. [I would tell myself] this will never happen again. We will never get trapped again."

For the first two years as a special operator, Tom was driven to be the best. And then, the realities of war shifted his worldview. From then on, Tom was driven by the desire to not die.

"Whatever breaks your ego or your thoughts of who you are and proves you wrong about yourself . . . that will break a man. I think that's what broke me. I thought I was invincible. [I'd think] *We are the best in the world . . . Oh, that kid just shot that guy.* So, there went that. *Oh, by the way, my friends are killing themselves at higher numbers than the enemy can kill us. So we're losing anyway.* I think realizing that you're not who you thought you were, and that the story you told yourself about those times [was off], I think that is what broke me."

In 2013, Tom came close to ending his own life. Had it not been for a text from Jen, the woman who would become his fourth wife, he would have joined the twenty-two veterans who die by suicide every day.

"You know, [by her] simply texting me, it saved my life, stopped me from killing myself. I [wouldn't talk to] a therapist unless they'd been to combat because they wouldn't get it. But [Jen] got it. She got it because she listened. She got it because she cared. And I nearly destroyed that. I almost destroyed the best thing I ever had, until I finally woke up. I mean it. Waking up with self-awareness for me was realizing [the pain] I was causing to other people."

For many combat soldiers, the moral injury to the psyche that comes from the burden of having to take life causes dark, complex mental and emotional responses, as does dealing with operator syndrome when back in the States.

"I haven't properly processed the shame for my actions during my job, because I think I still look at it as a job. But I realize the implications of it and the emotions attached to it now. And how I've merged that into my normal life and how I treated people, which I shouldn't have. The shame I feel has always come from my behavior at home and how I act, even when I know that behavior is not proper at home. That behavior is meant for combat, to keep you alive. And then, you know, Jen's understanding mind is always like, yeah, but you've been taught to be like that over there for so long, it's muscle memory, it's gonna happen at home. But I'm like, yeah, but I hate myself for this. I hate doing it. The brain is telling you, don't do it . . . and then you're doing it [anyway]."

DOING THE HARD THING

Tom admitted to Jamie that the hardest part about changing his responses and behavior from those of a warrior to a dad or husband was that he didn't *want* to stop being a warrior. He had to dig deep to find the courage to be someone else other than an operator, which was his deeply embedded identity, the part of himself that made him feel worthy and needed.

And, of course, there was the fact that he had a biological change to his sympathetic nervous system, and desire and willpower were not going to be enough to reroute those base, instinctual responses to a false danger that the amygdala was firing off.

"I've jumped up mad—though I'm telling myself, you know, you're just mad, this is stupid. Stop. But I don't wanna lose. So, I'm trying to win this argument. I'm trying to be right. I'm trying to be dominant and on top and win, instead of just saying it's okay and letting it go, because for so many years it's been: if you're not winning, you're dying. If you don't win, you're gonna be dead. But I'm learning now that I *won't* die. And by the way, the only thing that loses is our relationship when I'm trying to be right."

VIRAGO: THE FEMALE WARRIOR

When Jamie and his crew focused the camera on Jen Satterly, she tucked her long blonde hair behind her ear and then dug earnestly into the topic of PTSI, fueled by her passion for helping others.

She started out discussing how the same trauma symptoms she saw surfacing in Tom were recognizable in the other operators she'd worked with over the years. As director of film and photography of an elite Special Operations training company, Jen embedded with Navy SEALs, Green Berets, and Army Rangers in (RMTs) realistic military training exercises, sometimes for weeks at a time. "I've seen some shit," she said.

Jen also saw those symptoms in herself and realized she, too, had post-traumatic stress, due to a stressful childhood and a sexual assault in her teens. And like many long-term caregivers or soldiers' loved ones, she'd also developed secondary PTSI. Tom and Jen agree, Jen tends to physically "make herself small" during a disagreement, a sign of PTSI; something small is less likely to be chased or killed by a tiger.

Jen began researching PTSI and secondary PTSI and the symptoms. And the interventions. She and Tom decided they loved themselves as individuals enough that they would both do the hard things required to get healthy. They also decided they loved themselves as a couple so much that they would protect the relationship, treat it as an entity that they needed to work at to keep alive and healthy. Together, they were determined to come out on the other side of post-traumatic stress, and to make sure they were leaving a solid trail for others to follow.

The Satterlys cofounded the top-rated All Secure Foundation in August 2017 as a resource library to help warriors and their families navigate the maze of post-traumatic stress treatments and modalities. They worked individually with hundreds of warriors and their families and soon became more than a referral website but also a provider of education, awareness, and programs for healing. Both have written books about their journey.

That journey includes trying dozens of modalities in order to combat their PTSI and secondary PTSI symptoms. Some treatments have been more effective than others, but none with more than 50 percent efficacy or long-term lasting effects.

MOVING THROUGH THE WORLD, DIFFERENTLY

Eugene's procedure was a game changer for both of them. Jen's story of what happened next was powerful.

They decided to have the procedure done at the same time. "I was awake, looking at Tom lying on a patient bed in the next room. He'd been put under, using twilight anesthesia. Then Dr. Lipov put the shot in my neck; it wasn't pleasant but it wasn't terrible and it only lasted for about thirty seconds. Dr. Lipov bent over me, to tell me he was done and ask how I felt. I can't explain, I don't really have the words . . . there was a mix of euphoria, relief, sadness, and joy at the same time. It was so intense. Tears just started squirting out of me. The doctor rested his forehead on mine and said, 'I know, I know,' which made me cry harder."

Lying there, Jen felt lighter. But there was something missing. When she realized it was that the sense of waiting for a conflict or a confrontation was gone, this set off her tears all over again.

"After thirty minutes of sobbing, I laughed and told Dr. Lipov that he broke me. 'Welcome to repressed emotional experience,' he says. He put me at ease, he was emotional himself, he said he wished his mom and dad had been able to have the shot, that they might still be here."

Two weeks after the shot, she loved that she no longer felt she was walking on eggshells, that she no longer spent all her time dedicated to avoiding setting off Tom. The injections had made her feel empowered, able to do what was needed—or what was wanted—without worrying about backlash. She was experiencing joy again, reveling in happiness. She'd put the top down on the jeep, turn the music up, and sing at the top of her lungs when she drove down the highway. "I haven't been this free since high school, or had these feelings of hope and the fun of adventure. It was how I felt before I was raped."

She hadn't realized how much that loss of innocence and security had settled into her bones. "I felt like a kid again. One of the ways I really knew I'd experienced a change was that one of my odd Jen quirks disappeared. I'd had issues with eating in front of other people since being traumatized around this issue as a child. Tom and I were at a huge group buffet, I was starving and went first, filled up my plate, everyone watching me eat . . . I didn't even notice until Tom said something."

She grinned. "My timidity has faded; I'm less afraid, less of a pushover."

Tom's experience at Eugene's office is memorable for everyone who was there.

Coming out of the twilight after the injections, before he was fully conscious and aware of his current surroundings, a part of Tom's brain processed that he was in a room with medical equipment and came to the conclusion that he'd been captured and was about to be tortured. Thankfully, Tom had been restrained before the procedure, because his first reaction upon waking was fear and rage—he started thrashing against the restraints in order to get to his weapons, screaming at the shocked staff, "I'm going to kill every motherfucker in this room! You fu—Oh, hi, baby. Hi, Jen."

Jen had stepped into his line of vision, snapping him out of the delusion. Eugene claims this is one of his favorite party stories now, but it took a few hours for Tom, Jen, or the doctor to see the humor. To consider the amount of trauma Tom has endured to get him to that level of PTSI is horrifying.

The dual sympathetic reset is a godsend for people like Tom—and now it is available for everyday people, everywhere.

"After a few minutes, I realized this heavy blanket of worry, the feeling of *I'm late for something, something's wrong*. . . it was gone. I felt happy. I wasn't agitated. I could breathe, really breathe."

There was visual proof, not just anecdotal. The before and after scans show a remarkable change.

SPECT Scans

Patient #1

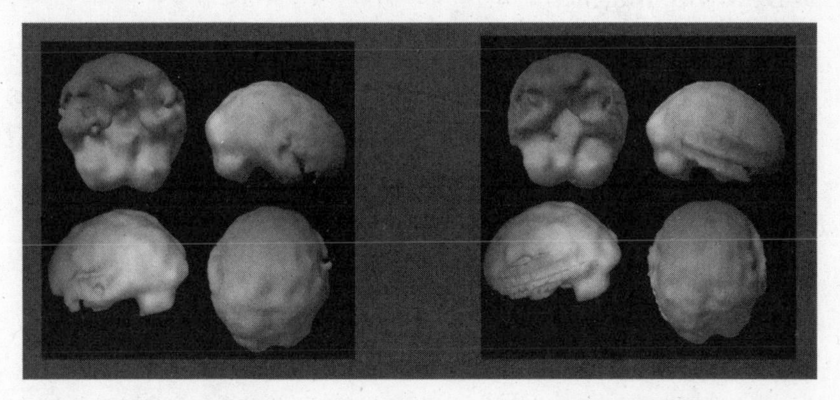

Before SGB **After SGB**

After the injections, Tom was refreshed, energized. While walking around Chicago a day later, Jen says, "Some dude ran up on us and Tom only said to him calmly, 'Step back, don't startle people.' That guy was lucky he didn't run into us a week ago."

Tom found he could do things out of routine, that he could be spontaneous, that he didn't freak out if he was walking the city streets and didn't know exactly where he was. He was able to be in a room full of people without developing anger, hyperanxiety, or fear. He suddenly had the ability to take a beat before reacting. There was a sense of peace. He was able to easily get up in the morning, feeling positive about what the day might hold, instead of dread and negativity.

The effects of DSR lasted for two months, until a group of young people next to them on a beach shot fireworks in their direction. Someone screamed to warn them. Jen saw Tom squat, arms out, ready for combat. Though an accident, the firework exploded within thirty feet of the Satterlys. That retriggered Tom's fight-or-flight mechanism. His PTSI symptoms have been less severe this time around, but he is scheduling another round of DSR with Eugene.

Tom was not deterred. "I've been happier than I've ever been. And now I know I can function like a normal human being. The stories I have heard from other warriors [after DSR] has been freaking remarkable. They can hold down jobs. Can get on planes. This is life changing."

Imagine what it can do for you.

CHAPTER TWELVE

The End Game

HISTORY THROUGH A NEW LENS

Once you have seen the effects of the procedure on people, the changes to their brain scans as well as how they carry themselves, it is hard to unsee it. If you take a few moments to consider pivotal times in history and the figures who led the pivots, also pause to think about how many of those historical figures (from dictators to school shooters) were likely dealing with years of trauma and stress with their invisible machine or sympathetic nervous system stuck in fight or flight and could have used a dose of DSR before making the irrational, anger-based decisions that affected large populations.

TAKE ACTION

So, what can you do if you or someone you know might have post-traumatic stress?

1. Go back to the PTSI questionnaire in chapter four. Or go back to chapter seven and read over the definition of ACEs. Do you see yourself in the descriptions? There are also a number of excellent resources listed in the appendix of this book (page 261).
2. If you think you may have PTSI, approach your doctor or a therapist. Remember, everyone has suffered trauma, so there is no shame in discussing yours with a professional. You will want to advocate for yourself and make sure you are being treated respectfully. Dr. Ochberg, in a 2021 interview with Jamie, is adamant on this point, saying, "I don't think we've done a good job with the post-traumatic stress injury definition when it comes to the shame component. Shame is something we overcome by admitting we have been ashamed." Trauma is not a weakness, or even a choice, so shame must be recognized for what it is in this instance—a product of societal stigma that needs to go the way of the dinosaur.
3. If the doctor is unsure about a PTSI diagnosis, go for a second opinion.
4. If there is a diagnosis, then discuss treatment options. We, of course, feel DSR is the soundest option for the majority of those with PTSI.
5. Stella Clinics can now be found around the United States (map on page 238) and in Australia and Israel, offering the DSR procedure under the guidance of Dr. Lipov and his team of doctors. For details and contact information, go to stellacenter.com/, or call (1-866) 497-9248.

6. DSR is not yet covered by insurance for most people in the United States. If finances are an issue, please visit Erase PTSD Now and apply for a grant at https://www.eraseptsdnow.org/home. Eugene and Erase PTSD Now are raising funds for those who need the procedure. If you would like to change lives by financially sponsoring a patient, please click on the link to donate. Erase PTSD Now is a 501(c)(3) nonprofit organization established in 2009; the organization is committed to sponsoring treatment, helping fund ongoing research, and promoting awareness about PTSD symptoms and treatment around the world. It's not about how much money the organization wants, it's about how many lives you can change.
7. Once you have had the DSR procedure, keep in mind you will still need to consider working on your emotional well-being. DSR resets your fight-or-flight mechanism, but it does not erase your memories or emotions, and it definitely does not resolve an existing problem that continues to cause trauma. And if you have TBI/CTE complications, you absolutely need to seek medical care specific to that diagnosis.
8. Spread the word.

THE HOPE

Trauma isn't going away. We're humans living on a planet rife with unavoidable complexities and strife. But the biological response to trauma that developed in us so long ago and then began working against us in the quiet moments . . . well, that is one problem that is avoidable. That is one problem that has been solved. The invisible machine can be fixed.

The tools are there now, thanks to Eugene and others.

"A holistic treatment that doesn't require mind-altering drugs may be the greatest scientific medical breakthrough since penicillin, maybe greater," says Jamie Mustard, "especially if you look at the volume of people affected and how it changes the way they and their families are now able to move through the world as themselves again."

Hopefully, those who are suffering from post-traumatic stress will utilize the tools and reset their invisible machine. Then, they can help shine a light of hope and helpfulness on the community in need around them. This is the way of change and progress.

Dr. Frank Ochberg states, "When you stand up for someone who can't stand up for himself or herself, you are doing the final act that we are privileged to do as a human beings. It is what is moral, honorable, difficult, but it means that we are carrying forth humanity and we should feel very good about that. I recognize that for a big part of my life I was achievement oriented and arrogant doing it. It's not that anymore. It's having the ability to do something that is going to be passed on."

So many of us have suffered.

Let's heal our wounds.

"What keeps me up at night? Boy, that's a loaded question," states Dr. Faber. "I'm sixty-two years old. There are so many things that I want to do to help make this world a better place before I go. And sometimes I just wish I had more time to do it. Whether it be to learn more, whether it be to help more with my youth foundation, or whether it is to help some of the homeless people who are in my neighborhood to get the care they need . . . there's a lot to do. And for me, there's just not as much time [as there once was]. And I'm left thinking on how I can maximize the maybe ten, twenty, thirty years I have left. I want to make this world a better place than when I first came in."

So many of the people in the pages of this book are striving and sacrificing for real change. Humans have always adapted and survived; we can do it again.

"My trauma gave me misery. My misery made me a scientist," states Eugene. "My hope is that everyone will know about the procedure and have easy access. My dream is to live to see Mercy Ships traveling the world and offering this to anyone who needs it, whether they can afford it or not. People don't need to hurt anymore, not like this."

LAST DOUBTS

Let us leave you with this:

If you have any doubts about the equanimity that is in your grasp, you should know that Green Beret Trevor Beaman recently called Jamie and said, "What would you think about offering DSR to my stepfather? Everyone deserves to get better."

Think about that. As a child, Trevor was continually sexually assaulted by his stepfather for years. As a Special Forces operator, Trevor dreamt up hundreds of ways to torture or kill the man. Now, after working with Eugene and getting the dual sympathetic reset, Trevor is able to take a step back from his pain and breathe. Jamie just listened as the Green Beret detailed how he isn't by any means looking to reconnect or have a relationship with the pedophile who hurt him for so long, but he has found that he no longer has murder and rage in his heart. Instead, he'd rather that a dark soul have all the tools available to make the choices that won't hurt others.

Trevor has suffered "big f-ing trauma," according to Dr. Lipov, on and off for most of his life. Most of us, thankfully, have not been chased by the same kind of tigers. But if someone who endured Trevor's level of trauma can find that kind of peace and grounding after DSR, imagine what it can do for the rest of us, no matter the scale of our stories.

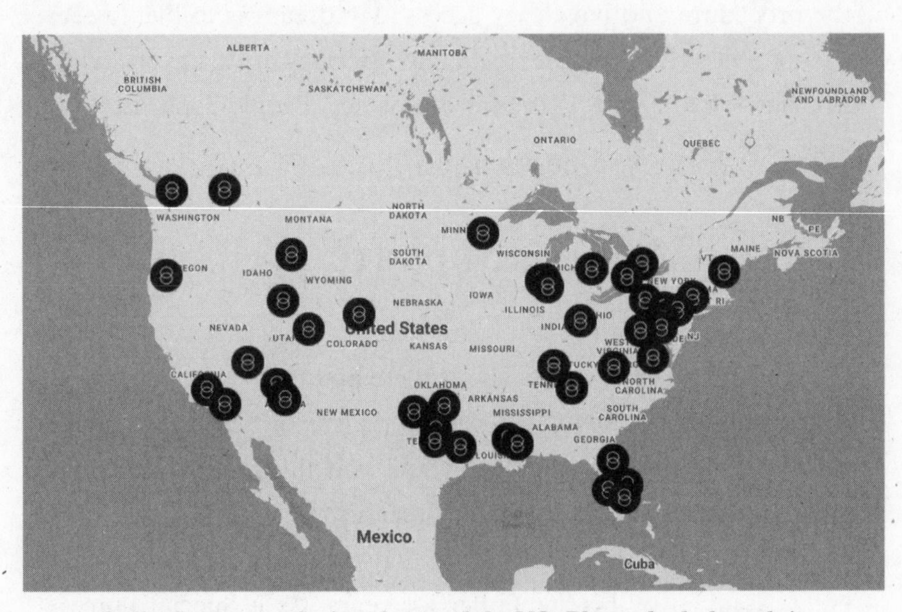

Stella Centers can be found around the US. Please check the website for updated and specific locations: www.stellacenter.com.

SPOKEN WORD POETRY

THE WAKE UP CALL

BY FRED MILES, 27 YEARS INCARCERATED

Education is the medicine

needed for our mental and spiritual poverty,

the restorative vision to our third eye

and to regain lost sovereignty.

In order to love one another,

we must first love ourself.

Hidden in these lines is a fountain of wealth.

I'm looking past your physical,

deep down into your heart and mind.

This is called

the true love of self, of kind.

The first law of the universe

is called self-preservation.

That means no hidden agendas

or self-destructive reservations.

The continuous cycle of Cain killing Abel

is only going to seal our fates.

Instead, we need to join together and elevate,

like two brain cell mates.
They got us striving for success
when in reality we fail.
They got us praying for heaven
while we livin' in hell.
The only cure for the struggle is the hustle.
It's called utilizing our collective brain muscles.
Now open your eyes.

There once was a magic power
in His spoken word.
He had the ability to resurrect the day,
a man once heard.
You know, we only love
their fully parted lips
and their goddess hips.
We once here vibrant men were Kings,
but now we've fallen into a state of a slave ship.
You see the art of love
and the science of sections.
One is sacred.

Now we killing our Black woman

and our daughters

and now they talking about Black lives matters.

Now, that is true Black-on-Black hate.

They say time

heal all wounds.

And that's our excuse to keep alive.

When our women be minimized

exploited in abuse.

Now all of this is called a meeting of the minds

and it's 360 degrees of knowledge,

wisdom, understanding for you to seek and find.

And if by chance it rang true,

it was because the spoken word was all about you.

Now open your eyes.

Our minds is the weapons,

shining brighter than a million suns.

Let's build new realities

and get the youth to put down the guns.

Know this:

Life ain't what it seems.

What if God was hiding

inside the shell

of a crackhead or a dope fiend?

What if there was no Malcolm,

Mandela,

or a Dr. King dreamin' his dreams?

You see, the duties of an elder

is to restore a lost truth,

to uplift our woman

and give knowledge and wisdom

and understanding to our youth.

They say the hand that rock the cradle

rules the world.

Now,

who runnin' the world?

Girls.

Thank you.

ACKNOWLEDGMENTS

I owe a debt of gratitude to the following people:

All the patients who have trusted me with their health and well-being. My wife and son for their support. The foundations providing amazing support for my dream of a society without PTSI, including the Glenn Greenberg and Linda Vester Foundation, Spookstock Foundation, Erase PTSD Now, and Tom and Jen Satterly's All Secure Foundation. Steven Taslitz of Sterling Partners. Our writer extraordinaire, Holly Lorincz. And, of course, my brother, Sergei Lipov, MD, for his amazing medical perspective.

—Eugene Lipov

Holly Lorincz for her tireless pursuit of the truth and the highest form of art.

Lieutenant Colonel Paul Toolan for his vision, action, and ability to get shit done.

Colonel Stuart Ferris for creating the opportunity and for his care and dedication.

Michael Humphrey for your belief, wisdom, and guidance.

Maryann Karinch, a soulful literary agent who has been a true friend and caring adviser. This book was her idea and I will never forget that. I believe it will save many lives.

Dr. Daniel Amen for his trust, belief, partnership, and paving the way.

Dr. Jay Faber for his profound empathy, brilliance, and making all things possible.

Dr. Frank Ochberg for his vision, brilliance, and being the first through the gate.

Dr. Gabor Maté for his HUMANITY.

Dr. Stephen Porges for his pursuit of scientific truth.

Sociobiologist and author Rebecca Costa for her foresight and unwavering conviction.

David Rae for his uncompromising and relentless pursuit of what's higher, and for giving me a chance.

Paul Bronkar for his prescience and talent.

Sean Keener for his intelligence, empathy, and getting me to ask the right questions.

Holt McCallany, a vulnerable juggernaut for art and social change.

Geoff Dardia for his knowledge, and relentless hard work toward the healing of his brothers.

Attorney Peter Vaughn Shaver who has been an invaluable protector and counselor.

Jared Nichols for his ability to see the future and put all the right people together.

Sheriff Alisa Gregory for her service, care, and compassion toward those who have been discarded.

Sara Harman for her truth and guts.

Hillary Bevan-Jones for her guidance.

Shaw and Michael Thomas for their faith and humanity.

Christine Toolan for her love of family and caring for all of us.

Corey Drayton
Trevor Beaman
Alex Pennacchia
Dr. Steven Nakana
Kevin Carroll
Fred Miles
Tom and Jenn Satterly
Steven and Denise Delellis
Nate and Katy Onniinniiwaat
Emmanuel Turaturanye
Dr. Carrie Elk
Julie Derrick
Arthur Derrick-Jones
Robbyn Peters Bennett
Donald Wilkes
Andy Spain
Will Marshall
Stephanie Marshall
Logan Marshall
Chuy Almonte
Paul Toolan Jr.
Ben Toolan
Matty Fiorenza
Elizabeth Fiorenza
Kevin Richmond
Melissa McPherson
Cecila Delacruz-Cenobio
Clarence Elkins
Molly Elkins
Roger Ledesma
Evalee Gertz
Meredith Greenberg
Linda Vester
Glenn Greenberg
Pippa Greenburg
Louie Psihoyos
Curt Smith
Antionette Young
Morris Adisanya Jr.
Alexis Smith
Nijel
Anna Kathrin Reid, for caring for Nijel
Brian Wannamaker
Jeffery Wannamaker
Michael Ulwelling
David Bentley

—Jamie Mustard

NOTES

1. Jamie Mustard. "Resident Iconist." *Advisors and Residents/ ForbesIgnite*. Accessed July 2022. www.forbesignite.com/about-us.
2. Eugene Lipov, MD. "Team." Stella Center. Accessed July 2022. www.stellacenter.com/.
3. National Council for Mental Wellbeing. "How to Manage Trauma (PDF)." National Council for Mental Wellbeing/Behavioral Health. Accessed June 2022. www.thenationalcouncil.org/wp-content/uploads/2021/04/Trauma-infographic.pdf.
4. Megan Hull. "PTSD Facts and Statistics." The Recovery Village. Last modified June 1, 2022. Accessed June 10, 2022. www.therecoveryvillage.com/mental-health/ptsd/related/ptsd-statistics/.
5. Stella Center locations and contact information can be found at www.stellacenter.com.
6. Sean Mulvaney, James H. Lynch, Matthew J. Hickey, et al. "Stellate Ganglion Block Used to Treat Symptoms Associated with Combat-Related Posttraumatic Stress Disorder: A Case Series of 166 Patients." *Military Medicine* 179, no. 10 (2014): 1133–40. doi.org/10.7205/MILMED-D-14-00151.

7. H. Wulf, MD and C. Maier, MD. "Complications and Side Effects of Stellate Ganglion Blockade. Results of a Questionnaire Survey." *Anesthetist* 41, no. 3 (1992): 146–51. German. PMID: 1570888. pubmed.ncbi.nlm.nih.gov/1570888/.
8. Dictionary.com, s.v. "Trauma." Accessed 2022. www.lexico.com/en/definition/trauma.
9. American Psychological Association. "Trauma." Accessed 2022. www.apa.org/topics/trauma.
10. American Psychological Association. "Trauma."
11. D. Janiri, MD; A. Carfì, MD; G. D. Kotzalidis, MD, PhD; et al. "Posttraumatic Stress Disorder in Patients After Severe COVID-19 Infection." *JAMA Psychiatry* 78, no. 5 (2021): 567–69. doi.org/10.1001/jamapsychiatry.2021.0109.
12. American Psychiatric Association. *Diagnostic and Statistical Manual of Mental Disorders,* 3rd rev. ed (DSM-III). (Washington, DC: American Psychiatric Association, 1987).
13. Military History Now. "A Brief History of Wartime PTSD: From Ancient Greece to Afghanistan." *Military History Now Blog.* September 17, 2012. Accessed May 2022. militaryhistorynow.com/2012/09/17/walking-wounded-ptsd-from-ancient-greece-to-afghanistan/.
14. Shakespeare. *Henry IV, pt. 1,* 2.3.59–61.
15. Edgar Jones. "Historical Approaches to Post-Combat Disorders." *Philosophical Transactions of the Royal Society B, Biological Sciences* 361, no. 1468 (March 24, 2006): 533–42. doi.org/10.1098/rstb.2006.1814.
16. T. Keller, MD, MPH and T. Chappell, MD. "The Rise and Fall of Erichsen's Disease (Railroad Spine)." *Spine* 21, no. 13 (1996): 1597–601. doi.org/10.1097/00007632-199607010-00022.
17. J. M. Da Costa, MD. "On Irritable Heart: A Clinical Study of a Form of Functional Cardiac Disorder and Its

Consequences." Reprint by *The American Journal of Medicine* 11, no. 5 (November 1951): 559–67. doi.org/10.1016/0002-9343(51)90038-1.

18. J. Satterly and H. Lorincz. *Arsenal of Hope: Tactics for Taking on PTSD, Together.* (Nashville: Post Hill Press, 2021), 16.
19. L. Abbott. "Art, Trauma, and PTSI: An Interview with Dr. Frank Ochberg." *A View from the Front/Journal of Military and Veterans' Health* 28, no. 3 (July 2020): 42–6. jmvh.org/article/art-trauma-and-ptsi-an-interview-with-dr-frank-ochberg/.
20. Eugene Lipov, MD. "Effects of Stellate-Ganglion Block on Hot Flushes and Night Awakenings in Survivors of Breast Cancer: A Pilot Study." *The Lancet: Oncology* 9, no. 6 (June 2008): 523–32. doi.org/10.1016/S1470-2045(08)70131-1.
21. Bonnie Miller Rubin and Tribune staff reporter, "Hot-Flash Sufferers Take a Shot at New Therapy." *Chicago Tribune.* August 10, 2005. https://www.chicagotribune.com/news/ct-xpm-2005-08-10-0508090306-story.html.
22. Eugene Lipov, MD, Sergei Lipov, MD, and Jamie T. Stark, MD. "Stellate Ganglion Blockade Provides Relief from Menopausal Hot Flashes: A Case Report Series." *Journal of Women's Health.* 14, no. 8 (October 19, 2005): 737–41. doi.org/10.1089/jwh.2005.14.737.
23. Eugene Lipov, Jaydeep R. Joshi, Sarah Sanders, et al. "A Unifying Theory Linking the Prolonged Efficacy of the Stellate Ganglion Block for the Treatment of Chronic Regional Pain Syndrome (CRPS), Hot Flashes, and Posttraumatic Stress Disorder (PTSD)." *Medical Hypotheses* 72, no. 6 (2009): 657–61. doi.org/10.1016/j.mehy.2009.01.009.
24. Frank Ochberg, MD. "An Injury, Not a Disorder." Dart Center for Journalism & Trauma. September 19, 2012. dartcenter.org/content/injury-not-disorder-0.

25. Lipov, Joshi, Sanders, et al. "A Unifying Theory Linking the Prolonged Efficacy of the Stellate Ganglion Block." 657–61.
26. Michael Alkire, MD; Michael Hollifield, MD; Rostam Khoshsar, MD; et al. "Neuroimaging Suggests That Stellate Ganglion Block Improves Post-Traumatic Stress Disorder (PTSD) Through an Amygdala Mediated Mechanism." The Anesthesiology Annual Meeting/ American Society of Anesthesiologists. October 24, 2015. asaabstracts.com/strands/asaabstracts/abstract.htm?year=2015&index=5&absnum=3003.
27. "Mental Health and Dissociative Fugue." WebMD. Reviewed September 27, 2020. www.webmd.com/mental-health/dissociative-fugue.
28. Jay Faber, MD. *Escape: Rehab Your Brain to Stay Out of the Legal System*. (CreateSpace, 2018).
29. Mulvaney, Lynch, Hickey, et al. "Stellate Ganglion Block Used to Treat Symptoms Associated with Combat-Related Posttraumatic Stress Disorder." 1133–40.
30. Alison Escalante. "Helping PTSD with a Shot: The New Treatments That Are Changing Lives." *Forbes*. February 2, 2021. www.forbes.com/sites/alisonescalante/2021/02/02/curing-ptsd-with-a-shot-the-new-treatments-that-are-changing-lives/?sh=3b5da9e69124.
31. R. Morey, MD; C. Haswell, S. Hooper; et al. "Amygdala, Hippocampus, and Ventral Medial Prefrontal Cortex Volumes Differ in Maltreated Youth with and without Chronic Posttraumatic Stress Disorder." *Neuropsychopharmacology* 41, no. 3 (2016): 791–801. doi.org/10.1038/npp.2015.205.
32. "About Marie Colvin." Marie Colvin Center for International Reporting/Stony Brook University School of Journalism. Accessed February 2022. mariecolvincenter.org/about-marie-colvin/.

33. Jared Nichols and Paul Toolan. "Ground Truth from Ground Zero Featuring Jamie Mustard." *The Best Pandemic Ever Podcast.* Podcast audio, September 21, 2020. podcasts.apple.com/ph/podcast/ground-truth-from-ground-zero-featuring-jamie-Jamie/id1525122107?i=1000491965261.
34. "Fort Bragg Army Base Guide." Base Guide/Military.com. Accessed February 2022. www.military.com/base-guide/fort-bragg.
35. B. Christopher Frueh, Alok Maden, J. Christopher Fowler, et al. "Operator Syndrome: A Unique Constellation of Medical and Behavioral Health-Care Needs of Military Special Operation Forces." *International Journal of Psychiatry in Medicine* 55, no. 4 (2020): 281–95. doi.org/10.1177/0091217420906659.
36. Frueh, Maden, Fowler, et al. "Operator Syndrome." 281–95.
37. John R. Blosnich, Melissa E. Dichter, Catherine Cerulli, et al. "Disparities in Adverse Childhood Experiences Among Individuals with a History of Military Service." *JAMA Psychiatry* 71, no. 9 (2014): 1041–48. doi.org/10.1001/jamapsychiatry.2014.724.
38. Lorincz, Satterly, *Arsenal of Hope: Tactics for Taking on PTSD, Together* (New York: Post Hill Press, 2021).
39. Rajeev Ramchand, Rena Rudavsky, Sean Grant, et al. "Prevalence of, Risk Factors for, and Consequences of Posttraumatic Stress Disorder and Other Mental Health Problems in Military Populations Deployed to Iraq and Afghanistan." *Current Psychiatry Reports* 17, no. 5 (2015): 37. doi.org/10.1007/s11920-015-0575-z.
40. Robbyn Peters Bennett, LPC. "Violence: A Family Tradition." YouTube video, 13:25, posted by TedXBellingham/TedXTalks. www.youtube.com/watch?v=WLMJHdySgE8&t=38s.
41. Fred Miles. "The Wake Up Call." Spoken poetry. The full text of the poem can be found on page 239.

42. Katherine Harmon. "Brain Injury Rate 7 Times Greater Among U.S. Prisoners." *Scientific American.* February 4, 2012. https://www.scientificamerican.com/article/traumatic-brain-injury-prison/?gclid=Cj0KCQjw39uYBhCLARIsAD_SzMSSYAkuqQtoxw6-xH5nkFOZrhkDB5UKFe2aMysX0evWqULIBMxhDAQaAv21EALw_wcB.
43. Eugene Lipov, Ryan Jacobs, Shauna Springer, et al. "Utility of Cervical Sympathetic Block in Treating Posttraumatic Stress Disorder in Multiple Cohorts: A Retrospective Analysis." *Pain Physician* 25, no. 1 (2022): 77–85. PMID: 35051147. https://pubmed.ncbi.nlm.nih.gov/35051147/.
44. Drue H. Barrett, Caroline Carney Doebbeling, David A. Schwartz, et al. "Posttraumatic Stress Disorder and Self-Reported Physical Health Status Among U.S. Military Personnel Serving During the Gulf War Period: A Population-Based Study." *Psychosomatics* 43, no. 3 (May–June 2002): 195–205. doi.org/10.1176/appi.psy.43.3.195.
45. Jennifer Berry. "What Are Neurotransmitters." *Human Biology/Medical News Today.* Reviewed/updated May 8, 2022. www.medicalnewstoday.com/articles/326649.
46. *Britannica*, s.v. "Scavenger Cell." Accessed January 2022. www.britannica.com/science/scavenger-cell.
47. Elizabeth Brododolo, PhD; Kahaema Byer, MS; Peter J. Gianaros, PhD; et al. "Stress and Health Disparities." *American Psychological Association Report.* 2017. www.apa.org/pi/health-equity/resources/stress-report.pdf.
48. Helen P. S. Wong, Judy W. C. Ho, Marcel Koo, et al. "Effects of Adrenaline in Human Colon Adenocarcinoma HT-29 Cells." *Life Sciences* 88, no. 25–26 (June 20, 2011): 1108–12. doi.org/10.1016/j.lfs.2011.04.007.

49. National Center for PTSD. "Posttraumatic Stress Disorder and Cardiovascular Disease (PDF)" *PTSD Research Quarterly* 20, no. 1 (2017). www.ptsd.va.gov/publications/rq_docs/V28N1.pdf.
50. National Center for PTSD, "Posttraumatic Stress Disorder and Cardiovascular Disease."
51. Kenneth A. Hirsch, MD, PhD. "Sexual Dysfunction in Male Operation Enduring Freedom/Operation Iraqi Freedom Patients with Severe Posttraumatic Stress Disorder." *Military Medicine* 174, no. 5 (2009): 520–22. doi.org/10.7205/MILMED-D-03-3508.
52. R. Yehuda, S. Southwick, E. L. Giller, et al. "Urinary Catecholamine Excretion and Severity of PTSD Symptoms in Vietnam Combat Veterans." *The Journal of Nervous and Mental Disease* 180, no. 5 (1992): 321–5. doi.org/10.1097/00005053-199205000-00006.
53. Claude M. Chemtob, R. W. Novaco, R. S. Hamada, et al. "Anger Regulation Deficits in Combat-Related Posttraumatic Stress Disorder." *Journal of Traumatic Stress* 10, no. 1 (1997): 17–36. doi: 10.1023/a:1024852228908.
54. Andrea Porzionato, Aron Emmi, Silvia Barbon, et al. "Sympathetic Activation: A Potential Link Between Comorbidities and COVID-19." *FEBS Journal* 287, no. 17 (2020): 3681–8. https://febs.onlinelibrary.wiley.com/doi/10.1111/febs.15481.
55. Luke D. Liu and Deborah L. Duricka, "Stellate Ganglion Block Reduces Symptoms of Long COVID: A Case Series." *Journal of Neuroimmunology* 362 (December 8, 2022): 577784, doi.org/10.1016/j.jneuroim.2021.577784.
56. Claudia Carmassi, Claudia Foghi, Valerio Dell'Oste, et al. "PTSD Symptoms in Healthcare Workers Facing the Three

Coronavirus Outbreaks: What Can We Expect After the COVID-19 Pandemic." *Psychiatry Research* 292 (October 2020). doi.org/10.1016/j.psychres.2020.113312.

57. "Post Traumatic Stress Disorder (PTSD) & Addiction: Signs, Symptoms & Treatment," American Addiction Centers. Updated July 13, 2022. americanaddictioncenters.org/co-occurring-disorders/ptsd-addiction.
58. Eugene Lipov, Maryam Navaie, Peter R. Brown, et al. "Stellate Ganglion Block Improves Refractory Posttraumatic Stress Disorder and Associated Memory Dysfunction: A Case Report and Systematic Literature Review." *Military Medicine* 178, no. 2 (2013): e260. https://doi.org/10.7205/MILMED-D-12-00290.
59. J. Davidson, D. Hughes, D. G. Blazer, et al. "Post-Traumatic Stress Disorder in the Community: An Epidemiological Study." *Psychological Medicine,* 21, no. 3 (1991): 713–21. doi.org/10.1017/S0033291700022352.
60. Verity Fox, Christina Dalman, Henrik Dal, et al. "Suicide Risk in People with Post-Traumatic Stress Disorder: A Cohort Study of 3.1 Million People in Sweden." *Journal of Affective Disorders* 279 (January 15, 2021): 609–16. doi.org/10.1016/j.jad.2020.10.009.
61. Holly C. Wilcox, Carla L. Storr, and Naomi Breslau, "Posttraumatic Stress Disorder and Suicide Attempts in a Community Sample of Urban American Young Adults." *Archives of General Psychiatry* 66, no. 3 (2009): 305–11. doi.org/10.1001/archgenpsychiatry.2008.557.
62. Justin Alino, Donald Kosatka, Brian McLean, et al. "Efficacy of Stellate Ganglion Block in the Treatment of Anxiety Symptoms from Combat-Related Posttraumatic Stress Disorder: A Case Series." *Military Medicine* 178, no. 4 (2013): e473. doi.org/10.7205/MILMED-D-12-00386.

63. Eugene Lipov, MD. "The Use of Stellate Ganglion Block in the Treatment of Panic/Anxiety Symptoms (Including Suicidal Ideation), with Combat-Related Posttraumatic Stress Disorder." In *Posttraumatic Stress Disorder and Related Diseases in Combat Veterans*, ed. E. Ritchie (New York: Springer, Cham. October 2015) doi.org/10.1007/978-3-319-22985-0_13.
64. Centers for Disease Control and Prevention (CDC). "Adverse Childhood Experiences Reported by Adults—Five States, 2009." *MMWR. Morbidity and Mortality Weekly Report* 59, no. 49 (2010): 1609–1613. https://www.cdc.gov/mmwr/preview/mmwrhtml/mm5949a1.htm.
65. Ellen Goldstein, MFT, PhD; Ninad Athale, MD; Andrés F. Sciolla, MD; et al. "Patient Preferences for Discussing Childhood Trauma in Primary Care." *The Permanente Journal* 21 (2017): 16–55. doi.org/10.7812/TPP/16-055.
66. ACE Interface. "Understanding Adverse Childhood Experiences: Building Self-Healing Communities." Accessed February 2022. www.sos.wa.gov/_assets/library/libraries/projects/earlylearning/understanding-aces-handout.pdf.
67. Office of the California Surgeon General. "Adverse Childhood Experiences (ACEs) and Toxic Stress." Updated 2022. osg.ca.gov/aces-and-toxic-stress/.
68. Office of the California Surgeon General. "Adverse Childhood Experiences (ACEs) and Toxic Stress."
69. Gary W. Evans, Jeanne Brooks-Gunn, and Pamela Kato Klebanov, "Stressing Out the Poor: Chronic Physiological Stress and the Income-Achievement Gap." Stanford Center on Poverty and Inequality. Accessed January 2022. cpi.stanford.edu/_media/pdf/pathways/winter_2011/PathwaysWinter11_Evans.pdf.
70. Morey, Haswell, Hooper, et al. "Amygdala, Hippocampus, and Ventral Medial Prefrontal Cortex Volumes."

71. David S. Moore, PhD. *The Developing Genome: An Introduction to Behavioral Epigenetics*, 1st ed. (Oxford: Oxford University Press, 2015).
72. Centers for Disease Control and Prevention (CDC). "What Is Epigenetics?" Genomics & Precision Health/CDC. Last reviewed August 15, 2022. www.cdc.gov/genomics/disease/epigenetics.htm.
73. T. Bothe, J. Jacob, C. Kröger, et al. "How Expensive Are Post-traumatic Stress Disorders? Estimating Incremental Health Care and Economic Costs on Anonymized Claims Data." *European Journal of Health Economics* 20, no. 6 (2020): 917–30. doi.org/10.1007/s10198-020-01184-x.
74. Transcript: *Finding of the Veterans' Disability Benefits Commission: Hearing before the Committee on Veterans' Affairs*, U.S. House of Representatives, 110th Cong., First Session. October 10, 2007. U.S. Government Printing Office, Washington, DC: 2008. www.govinfo.gov/content/pkg/CHRG-110hhrg39461/html/CHRG-110hhrg39461.htm.
75. Terri Tanielian and Lisa H. Jaycox, eds. *Invisible Wounds of War: Psychological and Cognitive Injuries, Their Consequences, and Services to Assist Recovery* (Santa Monica, CA: Rand Corporation, 2008), www.rand.org/pubs/monographs/MG720.html.
76. Matt Farwell and Elsa Givan. "Can a Single Injection Save Soldiers Suffering from PTSD?" *Playboy*. March 2016: 2–17. https://az601583.vo.msecnd.net/internal-documents/news/10375/2017/05/19/globalptsi_03_46_25.pdf.
77. Eugene Lipov, MD, and Ken Candido, MD. "Efficacy and Safety of Stellate Ganglion Block in Chronic Ulcerative Colitis." *World Journal of Gastroenterology* 23, no. 17 (2017): 3193–94. doi.org/10.3748/wjg.v23.i17.3193.
78. Kim Peterson, MS; Donald Bourne, BS; Johanna Anderson, MPH; et al. *Evidence Brief: Effectiveness of Stellate Ganglion*

Block for Treatment of PTSD (Washington, DC: Department of Veterans Affairs, 2017). https://www.ncbi.nlm.nih.gov/books/NBK442253/.

79. Lipov, Jacob, Springer, et al. "Utility of Cervical Sympathetic Block in Treating Posttraumatic Stress Disorder in Multiple Cohorts: A Retrospective Analysis."
80. Lipov, Jacob, Springer, et al. "Utility of Cervical Sympathetic Block in Treating Posttraumatic Stress Disorder in Multiple Cohorts: A Retrospective Analysis."
81. C. W. Hoge, MD. "Interventions for War-Related Posttraumatic Stress Disorder: Meeting Veterans Where They Are." *JAMA* 306 (2011): 549–51. doi:10.1001/jama.2011.1096.
82. Shankar Vedantam. "Nothing Proven for Treating Post-Traumatic Stress: Panel's Findings Attract Attention as More Troops Are Affected by Disorder." *Houston Chronicle*. October 19, 2007. www.pressreader.com/usa/houston-chronicle/20071019/281663955645067.
83. Melissa Wesner, LCPC. "Brainspotting vs. EMDR: How They're Similar and How They're Different." Lifespring Counseling Services. Accessed January 11, 2021. lifespringcounseling.net/brainspotting-vs-emdr/2021/1/11/brainspotting-vs-emdr-how-theyre-similar-amp-how-theyre-different.
84. Eugene Lipov, MD, and Briana Marie Kelzenberg, MS, PC-A, *Exit Strategy for Posttraumatic Stress Disorder* (Self-published. 2011) 59–60.
85. Jan E. Kennedy, Michael S. Jaffee, Gregory A. Leskin, et al. "Posttraumatic Stress Disorder and Posttraumatic Stress Disorder-Like Symptoms and Mild Traumatic Brain Injury." *Journal of Rehabilitation Research and Development* 44, no. 7 (2007): 895–920. doi.org/10.1682/jrrd.2006.12.0166.
86. Alexis Peterson, Likang Xu, Jill Daugherty, et al. "Surveillance Report of Traumatic Brain Injury-Related Emergency Dept Visits, Hospitalizations, and Deaths—United States, 2014."

Centers for Disease Control and Prevention (2019). Accessed March 2022. https://www.cdc.gov/traumaticbraininjury/pdf/TBI-Surveillance-Report-FINAL_508.pdf.

87. Research!America. "Traumatic Brain Injury." www.researchamerica.org/sites/default/files/TBI_0120.pdf.
88. A pseudonym. "Rose" was interviewed by the authors in February 2022.
89. Sara was interviewed by the authors in February 2022.
90. Summary by World of Work Project. "Drowning Rats Psychology Experiment: Resilience and the Power of Hope." Accessed November 2022. worldofwork.io/2019/07/drowning-rats-psychology-experiments/.
91. Rubin and Tribune staff reporter, "Hot-Flash Sufferers Take a Shot at New Therapy."
92. Steven R. Hanling, Anita Hickey, Ivan Lesnik, et al. "Stellate Ganglion Block for the Treatment of Posttraumatic Stress Disorder. A Randomized, Double-Blind, Controlled Trial." *Regional Anesthesia and Pain Medicine* 41, no. 4 (2016): 494–500. rapm.bmj.com/content/41/4/494.
93. Peterson, Bourne, Anderson, et al. *Evidence Brief: Effectiveness of Stellate Ganglion Block for Treatment of Posttraumatic Stress Disorder (PTSD).*
94. Eugene Lipov, MD. "Letter to the Editor: Stellate Ganglion Block for Posttraumatic Stress Disorder: A Call for the Complete Story and Continued Research." *ASRA*. July 19, 2018. www.asra.com/news-publications/asra-newsletter/newsletter-item/asra-news/2018/07/19/to-the-editor-stellate-ganglion-block-for-posttraumatic-stress-disorder-a-call-for-the-complete-story-and-continued-research.
95. Kristine L. Rae Olmsted, Michael Bartoszek, Sean Mulvaney, et al. "Effect of Stellate Ganglion Block Treatment on

Posttraumatic Stress Disorder Symptoms: A Randomized Clinical Trial." *JAMA Psychiatry* 77, no. 2 (February 2020): 130–8. doi.org/10.1001/jamapsychiatry.2019.3474.

96. Michael Gier. *Wounded Heroes*. Gier Productions Documentary. 2021. www.woundedheroesdocumentary.com/.
97. "PTSD Treatment Helped Two Men Heal from Trauma of Gun Violence." *The Doctors*. CBS. Aired May 21, 2021. www.thedoctorstv.com/videos/ptsd-treatment-helped-two-men-heal-from-trauma-of-gun-violence.
98. Heather Abbott and Matthew Polevoy. "New Army-Funded Research Shows Promise for PTSD treatment." CBS NEWS/*60 Minutes/Overtime* segment. Aired November 6, 2019.
99. James H. Lynch, Peter D. Muench 2, John C. Okiishi, et al. "Behavioral Health Clinicians Endorse Stellate Ganglion Block as a Valuable Intervention in the Treatment of Trauma-Related Disorders." *Journal of Investigative Medicine* 69, no. 5 (2021): 989–93. doi.org/10.1136/jim-2020-001693.
100. Athlete Concussion Foundation: athleteconcussions.org/.

RESOURCES

For immediate crisis situations and assistance:

For immediate assistance, call 988 (English- and Spanish-language capabilities) or go to your closest hospital emergency room. It is critical to be honest and clear about the severity of symptoms, especially thoughts of suicide.

HOTLINES and HELPLINES (in alphabetical order):

- COPLINE—Provides complete confidentiality as well as anonymity if the caller chooses. Staffed by retired officers. (1-800) COPLINE (267-5463).
- SAMHSA's Disaster Distress Helpline—24-7, 365-day-a-year crisis counseling and support to people experiencing emotional distress related to natural or human-caused disasters. Call or text (1-800) 985-5990.
- FIRE/EMS Helpline—(1-888) 731-FIRE (3473).
- National Domestic Violence Hotline—24-7 support in English, Spanish, and 200-plus languages through interpretation service. (1-800) 799-7233.

- National Human Trafficking Hotline—(1-888) 373-7888.
- National Sexual Assault Hotline—(1-800) 656-HOPE (4673).

Support for PTSI

- Stella (for stellate ganglion block, ketamine, and other cutting-edge care). Stella has a nationwide network of clinics that provide cutting-edge treatment for post-traumatic stress symptoms. Stella is based on the understanding that trauma causes a biological injury, which can be healed with the right treatments and the right support. Stella provides treatments like dual sympathetic reset and ketamine. When Stella treatments are used, trauma sufferers can achieve long-lasting healing from trauma.
- Erase PTSD Now is a 501(c)(3) that sponsors DSR procedures and other Stella Clinic treatments, helps fund research, and promotes awareness of PTSI.
- In partnership with community health organizations, the Trauma Foundation's Trauma Resilience Initiative connects best-practice trauma clinicians with underserved populations lacking the resources to access appropriate support and treatment. This initiative facilitates and funds participants to receive at least six months of individual treatment involving weekly sessions of regular therapy, individualized to support their personal development of resilience and regulation.
- Give an Hour is a network of volunteer counselors, therapists, and other licensed professionals. They provide free mental health counseling to victims of mass violence,

survivors of interpersonal violence, rare-disease caregivers, and combat veterans returning from Iraq and Afghanistan, and they engage in opioid crisis response.

- Vets4Warriors is a nonprofit, founded by and run by veterans, dedicated to military mental health support and suicide prevention. Find them online or call (1-855) 838-8255 for a 24-7 helpline for peer support and resources.
- Blue H.E.L.P.'s mission is to reduce mental health stigma through education, advocate for benefits for those suffering from post-traumatic stress, acknowledge the service and sacrifice of law enforcement officers we've lost to suicide, assist officers in their search for healing, and bring awareness to suicide and mental health issues. Email contact@bluehelp.org.
- First Responder Support Network's mission is to provide educational treatment programs to promote recovery from stress and critical incidents experienced by first responders and their families. Email info@frsn.org.
- OSI CANADA (targeted to first responders in Canada) does not see post-traumatic stress as a disorder, they see it as an injury. If you are a first responder who is suffering from the symptoms of an occupational or operational stress injury, a PTSD or PTSI diagnosis is not required to get their help.
- The Rape, Abuse & Incest National Network (RAINN) is an American nonprofit anti–sexual assault organization, the largest in the United States. RAINN operates the National Sexual Assault Hotline, as well as the Department of Defense Safe Helpline, and carries out

programs to prevent sexual assault, help survivors, and ensure that perpetrators are brought to justice through victim services, public education, public policy, and consulting services. Go to www.rainn.org.

- Network for Victim Recovery of DC empowers victims of all crimes to achieve survivor-defined justice through a collaborative continuum of advocacy, case management, and legal services. Go to www.nvrdc.org.
- ACEs Aware Questionnaire (Adverse Childhood Experiences Revised Questionnaire) can be found at acesaware.org.
- National Center on Domestic Violence, Trauma, and Mental Health collects information about groups and organizations that support those who have suffered from domestic violence. Go to www.nationalcenterdvtraumamh.org/resources/national-domestic-violence-organizations/.

INDEX

INDEX

ABOUT THE AUTHORS

Dr. Eugene Lipov is a complex anesthesiologist and has been called the "Einstein of modern anesthesiology." His discovery and innovation, the Dual Sympathetic Reset (DSR), was endorsed by President Obama in 2010. His research has an 85 to 90 percent success rate in reducing the effects of trauma.

Jamie Mustard is an artist, futurist, multi-media consultant, and writer, including his work on perception in the physical world. As an Iconist, his passion is to teach the science and art of obviousness, the anatomy of what causes any idea, art, or message to stand out and take hold.

ABOUT THE WRITER

Holly Lorincz is a successful collaborative writer and the owner of Lorincz Literary Services. She is an award-winning novelist (*Smart Mouth*, *The Everything Girl*) and co-writer (including the bestselling *Crown Heights*, *How to Survive a Day in Prison*, and *Arsenal of Hope*), a nationally recognized speaking coach, and the proud new proprietor of Cloud and Leaf Bookstore on the Oregon coast.